Routledge Revivals

Co-operative Industry

Ernest Aves (1857-1917) was an influential social analyst and civil servant. This title, first published in 1907, during Aves' work for the Board of Trade, investigates the different forms of industrial co-operation within Britain; the fundamental principle of this is arguably "equitable association", leading to increased profitability and the strengthening of industry. Chapters discuss such areas as centralisation, co-operative production and co-operative agriculture. This interesting reissue will be of particular value to students of economics with an interest in co-operative industry and the history of economic thought.

Co-operative Industry

Ernest Aves

ROUTLEDGE REVIVALS
Routledge
Taylor & Francis Group

First published in 1907
by Methuen & Co

This edition first published in 2014 by Routledge
2 Park Square, Milton Park, Abingdon, Oxon, OX14 4RN

Simultaneously published in the USA and Canada
by Routledge
711 Third Avenue, New York, NY 10017

Routledge is an imprint of the Taylor & Francis Group, an informa business

Publisher's Note
The publisher has gone to great lengths to ensure the quality of this reprint but points out that some imperfections in the original copies may be apparent.

Disclaimer
The publisher has made every effort to trace copyright holders and welcomes correspondence from those they have been unable to contact.

A Library of Congress record exists under LC control number: 07041769

ISBN 13: 978-0-415-73451-6 (hbk)
ISBN 13: 978-1-315-81993-8 (ebk)

CO-OPERATIVE INDUSTRY

BY

ERNEST AVES

METHUEN & CO.
36 ESSEX STREET W.C.
LONDON

First Published in 1907

PREFACE

FROM time to time the pressure of other work has led to the postponement of the actual completion of this little volume, which had been almost written by January, 1905.

In its production I have been placed under obligations to many, but especially to Mr. J. J. Dent, of the Board of Trade, who has not only given me the benefit of his long, intimate, and exceptional knowledge of the Co-operative Movement, but has, with very great kindness, undertaken to bring figures, when necessary, up to date, and to see the volume through the press during my absence from England.

E. A.

OFF PORT ADELAIDE

April 1907

CONTENTS

CHAPTER VIII

PART II.—THE WORKSHOP

CHAPTER IX

CHAPTER X

CHAPTER XI

CHAPTER XII

CHAPTER XIII

CO-OPERATIVE INDUSTRY

INTRODUCTION

General Divisions of the Subject—The Beginnings of Industrial Co-operation—Robert Owen—The "Industrial Revolution"—Economic Inter-dependence—The Rochdale Pioneers—The Christian Socialists—Historical Landmarks—National Congresses—The Co-operative Union—Labour Co-Partnership Association—The Productive Federation—The International Co-operative Alliance—Agricultural Organization Societies—Women's Guilds—Past and Present

INDUSTRIAL Co-operation is found in many forms, and in many degrees of completeness. The working-class "store," the "Productive" workshop or factory, the "Profit-sharing" Joint Stock Company or private firm, the varieties of associated enterprise in agriculture are among its numerous types. Most of them could be again subdivided, and it would be useless to attempt to describe them all by any single formula. Constitution, financial basis, aim—all vary, and in consequence the part that they play alike in the co-operative movement itself, and in the industrial life of the nation.

The wider diffusion of profits, the maintenance of industrial peace, increased " efficiency," the strengthening of a threatened industry of prime national importance are among the ideas that are apt to assume special prominence in connection with one or another of these different co-operative forms, and thus while the significance of all may be included under some such general national aim as that of "increasing economic and social well-being," no narrower statement would suffice. Since, therefore, co-operation stands for many things, it will be necessary to divide up the subject-matter of this volume, and the earlier chapters will be concerned with working-class co-operation, and its exceptional development in this country in the shape of the store and of the industrial and social movement of which the store is the most widely diffused, and most obvious sign. " Productive" co-operation and the partial application of co-operative principles to non-co-operative business and to agriculture, will form the subject-matter of later sections.

The cardinal principle of industrial co-operation is that of "equitable association," and in this there is nothing that is either new or even modern. In some of its applications, indeed, the principles of co-operation go deep down into the more ordinary forms of industrial life. They are peculiar to no epoch and to no people, and the independent productive society itself is, as Professor Marshall pointed out in his inaugural address at the Co-operative Congress, in 1889, "representative of a very ancient race."

Although, however, origins may thus be lost in a distant retrospect, the beginnings of industrial co-operation as we know it in this country to-day are rightly traced to much more modern sources, partly personal—an impulse springing from a single life, and partly historical—which perhaps is only another way of saying that the engendering impulses were diffused among large masses of the people. As regards the former, the source of inspiration is found in the life of one strenuous and famous man, Robert Owen, while those who first laid the sure foundations of success were a patient group of comparatively unknown men—the Rochdale "Pioneers."

As regards the wider impersonal forces, they are found in the changes wrought by that industrial revolution which, long before the little group of twenty-eight men in Rochdale had with difficulty collected their twenty-eight pounds of capital and started their little store, had swept away for ever the old peaceful but stagnant domestic system of industry, and in introducing the "large system," had at the same time created the new class, numerous and distinct, of the factory employee. Much of the history of co-operation is the record of an industrial expedient by which this new class has endeavoured to deal with the new problems, social and economic, with which it was confronted.

In the changes and in the industrial difficulties and impulses of the early years of the nineteenth century, the chief explanation of the beginnings of industrial co-operation are thus found. They were times of stress, when food was dear, and wages low, when legal disabilities hampered labour at every turn, when women and children were unprotected by the Factory Acts ; and when home conditions were being deteriorated and home stability threatened by the pressure of the new economic conditions that mechanical inventions and the rapid growth of the manufacturing industries were introducing. Wealth was being rapidly amassed, but in spite of this, the spread of well-being for large sections of the people was checked. Population was rapidly increasing, but for many the country did not produce the means of mere subsistence, and it was between 1795 and 1834 that, as Arnold Toynbee has said, "the problem of pauperism came upon men in its most terrible form."

It was a time that had made Cobbett an agitator, and that led Lord Shaftesbury to become a social reformer; Carlyle to write his "Past and Present," and Mrs. Browning to chant her "Cry of the Children." The Luddite Riots of 1811 had been an episode not less of suffering than of ignorance, and the political forces of working men that had helped to bring about Reform in 1832 gathered again in a fever of disappointment in the Chartist Movement of 1839.

The experiences of these early decades were thus object-lessons of social and industrial needs, and pressed with unusual force upon the sympathetic imaginations of thoughtful men. No mind was more responsive than that of Robert Owen himself, an enlightened employer possessed of unbounded energy, and from an early period of his life, as he had described himself, "an active man of business, of close and accurate observation." It was this man, who had started business life as a lad, and had become the manager and part proprietor of mills at nineteen, who was led to re-model the conditions of employment of his own work-people, and was thus induced by his great success, to experiment with wider projects of reform, to elaborate his "rational system" of a new order of society, and to expound the doctrine of a communistic basis of life in which men and women brought under new educational influences were to work with a new social impulse, in which remuneration should no longer be based upon a competitive wage system, and in which the stress, the uncertainty, the profit seeking, and the harmful fluctuations of the open market were to be abolished.

Owen showed himself in the earlier years of his life to be something of a practical genius, but in his teaching he proved himself to be something more. He was an idealist and an optimist, and, perhaps, by the very virtues of his mind, was led to place too ready a confidence in the power to realize a dream of industrial life in which men and women inspired to "temperate but effective labour," by "a community of mutual and combined interest," should afford a practical demonstration of the possibility of eliminating the worst evils of distress and unemployment.

Owen, like so many other reformers, did not allow sufficiently either for the inferiority, the apathy, or the ignorance of the individual units upon whom, in the last resource, every social public reform is dependent, or for the slow workings of intellectual and moral change. Neither, it must be added, did he allow for the complexity of industrial life, or realize its essentially organic structure. His schemes were often at once too simple, too artificial, and too optimistic, and though they were based on the hopeful anticipation of a widespread personal reform, and intended to be democratic in their constitution and conception, they would have required in practice the guiding presence of a moral if not of an administrative autocrat.

Although, however, his proposals led to no prolonged successful concrete effort, yet, from their ethical attractiveness, and because of the seriousness of the evils from which they were designed to help men to escape, they stirred the generous impulses and the hopes of large numbers of the people. Owen, forceful in action and full of ideas, had become, as Sir Leslie Stephen has written, "one of the most important figures in the social history of the time," and many were ready to follow his inspiring lead, and to make serious efforts to give effect to his proposals.

For these, as for ordinary industrial enterprises, large supplies of capital were necessary, and it was thus mainly as instruments by which capital could be acquired to be used for communistic purposes, that from 1827 to 1832, the Owenites founded large numbers of co-operative societies, between four and five hundred having, it is said, been in existence in the last-mentioned year. The following extract *

* Quoted in Benjamin Jones' "Co-operative Production," Vol. I., p. 77.

from the *Brighton Co-operator*, of May, 1828, will sufficiently explain the hopes and aims of these early efforts.

"Wherever you go," the editor wrote, "you hear of hard work, low wages, and pauperism. This distress is the inevitable consequence of working for others instead of working for themselves. . . .

"It is capital we want. . . . We must form ourselves into a society for this special purpose; we must form a fund by weekly deposits; as soon as it is large enough, we must lay it out in various commodities, which we must place in a common store, from which all members must purchase their common necessaries, and the profit will form a common capital to be again laid out in the commodities most wanted. Then we shall have two sources of accumulation—the weekly subscription and the profit. . . . The society will be able now to find work for some of its own members, the whole produce of whose labours will be common property. . . . As the capital accumulates still farther, it will employ all the members, and then the advantages will be considerable indeed. When the society has accumulated sufficiently, the society may purchase land, live upon it, cultivate it themselves, and produce any manufactures they please, and so provide for all their wants of food, clothing, and houses. The society will then be called a community. . . . But if the members choose to remain in a town, instead of going into a community, they may derive all the advantages from the society which I have stated."

As if for settlers in a new country, the picture is drawn; with a little capital and a little land, it was hoped that a self-contained community could be formed, of those who, withdrawing if need be "from the town," could provide "for all their wants." The extract reflects the widespread fallacy that it was easy to provide a way of escape from an industrial environment from which escape was for all most difficult, and for the great mass of men, impossible. And the difficulty increases. Administrative reform, largely sanitary, and the ways in which government, local or central, spreads its tentacles, alone tend to make the general life more organic, the individual less entirely free. Moreover, the nerve strands of industrial life are far more numerous now than then, their network more interwoven and more widely spread. The industrial world, at the same time, grows smaller as well as more closely knit together, and as distance is annihilated and the powers of intercommunication are increased, the inter-dependence of every part becomes an increasingly important and real fact. This is at once the source of the hope of industrial stability in the future, and the cause of many of the disturbing fluctuations of the present time, but it leaves uppermost in our minds the idea of healthy development rather than that of isolation, on however lofty a pinnacle, or of re-construction, however carefully it be planned. Thus, save in exceptional and insignificant forms, it leaves the idea of self-contained communistic life, be it by units or by groups, more and more of a chimera every day.

It follows, therefore, that however idealistic and however considerable the moral worth of separatist industrial schemes to-day may be, none have great practical significance or even importance. The great working-class movements of to-day in this country—co-operative, trade union, and friendly society—are thus found, each in its own way, trying either to fashion an industrial life that in its main features and foundations is accepted, or, more often, simply to strengthen the position of those who stolidly take their part in it. The modern co-operator is, it is true, more than any other, trying to carve out from the whole a sphere that shall be determined by his own scheme of things; but even with him, however bold his dreams of the future may be, the vastly greater part of his effort is conducted under conditions which are, for the most part, acquiescent in those under which the main body of the industry of the country is carried on.

Moreover, even as co-operative industry spreads, it becomes, at the same time, more and more complex and more dependent. Factories, for instance, are built, and the most modern machinery, made often by individualistic patentees, is introduced; or housing schemes are entered upon, and the manifold products and conveniences of the world of non-co-operative manufacture are freely requisitioned. Sectional independence is, or should be, a bygone cult driven out by the mighty forces that make constantly for a great unity. It is realized, for instance, that the only hope of stability for even such a modern project as that of a "Garden City"—so attractive and even so hopeful in many of its aspects—must be found in the presence of either manufacturers or individuals who, for the most part, find their market or their means of livelihood in the greater world outside.

In the same way, the communistic principle of equality of remuneration has been beaten out of court by the inequality of powers and the varying gifts of energy and will possessed by men, and the demands for greater "equality of opportunity" and for a "living wage" are its nearest modern equivalents. Finally, it has been found impossible to dispense with the impelling and guiding force of individual profit in industry.

It would, however, be misleading and unjust to Owen's memory to conclude that because "Owenism," as embodied in concrete projects of industrial reform, led to no great successes, his is a record of failure. On the contrary he, perhaps more than any other man, sowed the seed and tilled the stubborn soil out of which many of the most fruitful efforts, administrative, educational, and industrial, were destined to spring. In the strikingly successful management of his own business, he demonstrated the practical truth of the principle that, even in their own interests, employers should consider the welfare of their workpeople. The Co-operative store that he himself established, and which has been described, so fertile was he in experiment and projects of reform, as "a mere detail" in the management of his factory, has helped to pave the way for that distributive store movement which is to-day the most conspicuous form that co-operation takes in this country. Further, Owen looked always above all else for the introduction and spread of political action, of an industrial system and of social ideas that should be productive of well-being and formative of character, and in the Factory Acts, in the now accepted principle that, if men are to have a fair chance of living decent lives, the material conditions by which they are surrounded must be also decent, above all in our national system of education, and in the growing practical recognition of the fact that in the training of life when young is found the best guarantee of trained and disciplined life when the more formative years have passed by, his influence is seen. He thus takes his place in history as a pioneer of the first order—of the order whose greatest achievements are found not in what they accomplish, but in what they initiate, in what they point to not less than in what they reach. Although, as schemes, with the common weakness of most schemes—the too rigid application of theoretical conceptions—many of Owen's plans broke down, the value of much of his teachings remains for this generation hardly less than for his own to assimilate—the lessons of the sense of communal service, of the more generous recognition of the greater equality of claims and hopes that all men in the absence of restrictive and maleficent training should be able to share, and, if wealth be amassed, the acceptance of its great responsibilities no less than its delights.

In other directions, without any of the ulterior objects of the Owenites in view, co-operative societies were also being formed in these early years, much as they often are to-day, "as a defence against the inroads of the distributing classes on the working-men's pocket; and also as a means of promoting ready-money dealings, and the prudence in expenditure which usually accompanies such dealings." *

The conditions under which ordinary retail trading was carried on pointed to the need of such societies. "The back streets of the manufacturing towns swarmed," we are told, "with small shops in which the worst of everything was sold, with unchecked measures and unproved weights." † But the difficulties were great: improvidence was rife, and "the general indebtedness among the working people made success almost impossible."

Useful though any new plan of retail distribution was, that helped to check the current practices on the one hand of offering inferior commodities at extravagant prices, and on the other of reckless spending, the early form of co-operative store had no special attraction for the industrial classes. From the modern co-operative point of view, the economic basis was unsound: "a certain number of persons

* Ludlow and Lloyd Jones, "Progress of the Working Classes, 1832–1867," p. 132.

† Ibid.

supplied the capital in small shares; and divided in proportion to the capital invested whatever profit was made." In some cases capital was put in in larger sums by the well-to-do and the well-disposed; but the constitution of the modern co-operative store, although not entirely unknown, had as yet attracted no attention, and its adoption had been attended by no marked successes.

It does not come within the scope of this little volume to attempt to write even the co-operative chapter of economic history. This has been already done, sometimes in very full detail, and sometimes in brilliant outline, by more than one writer; but it is desirable to turn the mind of the reader for a moment to a few landmarks in co-operative history, and especially to the work of the Rochdale Pioneers, who have been already mentioned.

It is because of the success of their venture, that the Pioneers are regarded as the practical founders of the co-operative store movement in England. They made no great discovery, but they put their enterprise on a sound, and, for those to whose suffrages they afterwards appealed, on an attractive basis.

The first item of their programme was "the establishment of a store for the sale of provisions, clothing, etc.," but they hoped also to build homes for some of their members; to manufacture articles so as not only to produce commodities that their members might acquire, but to provide "employment for such members as may be without employment, or who may be suffering in consequence of repeated reductions in their wages." Unemployment and low pay were also to be met by the provision of employment upon the land, and it was hoped that after a time the society might be able "to arrange the power of production, distribution, education, and government" for a "self-supporting home colony of united interests" and to help other societies to establish such colonies.

Hopes and aims were therefore of the widest, but the practical success of these men is always identified with their comparatively humble object of "establishing a store." The essential features of this part of their scheme were the payment of a fixed rate of interest on the share capital, and dividing up the elastic element of profits among the members according to the amount of their purchases. Except in this form individual profit disappeared from their scheme, and its place was taken by the now celebrated co-operative "dividend." The vexed question as to whether the employee as such was entitled to any share in the profits was probably hardly considered, for it is one of those questions that only assumes importance when some measure of success is being achieved, when organization is somewhat elaborate and operations are extensive. In the early days of the Pioneers, with a total capital of £28 to deal with that had been collected among themselves through the slow process of twopenny or threepenny subscriptions per week, and with a small store opened for the supply of "a few common articles for the consumption of their own families," it is known that at first all service was voluntary. It is certain that no vision could have been expected to see in future developments the importance that has since attached to this particular problem of remuneration.

Combined with business management of average capacity, the secret of at least the financial prosperity of the co-operative movement is found in the adoption of the two principles that have been mentioned: the fixed rate of interest on capital, and the payment of dividend to members on the amount of their purchases. Membership of a co-operative society on this basis becomes a real thing, and adopting, as the stores do, the political principle of representative government, the whole spirit and aim in business management is at once democratized. All inducement to buy inferior articles, to charge high prices, to adulterate, or to give short weight or measure should disappear, for the only effect of taking some unfair advantage of the member as customer or consumer, would be to swell the margin of profit that would go back to him afterwards in the shape of increased dividend. Thus, the natural preference for fair dealing that all honest men share, whatever part they play in industrial life, was fortified not only remotely, but directly and immediately by considerations of self-interest. At every stage—in buying stock, in selling, and in sharing results, the basis of a real corporate interest was securely laid.

Upon this basis, to a consideration of which we shall have to return, an imposing structure has been raised during the last sixty years, the total sales of the stores, founded on the Rochdale model, having reached a total of 61 millions in 1905, and their membership nearly 2¼ millions.

A few other landmarks in the co-operative movement may be mentioned, either as incidents in its development or as national events that have had an important bearing upon it.

Two years after the Rochdale Pioneers started in 1844, the Corn Laws were repealed. In 1847 the Ten Hours Act was passed, and women and children thereby directly, and many men indirectly, protected against the excessive working hours of the past, securing thus that increased leisure, the possession of which is almost a condition of individual progress. Fresh legislative steps were also taken towards giving freer scope to the great principles of industrial combination and association, and in 1848 came the hopeful wave of revolutionary ardour.

The Workmen's Associations started then in France were watched by Englishmen, and, through Mr. J. M. Ludlow, became an incentive to the Christian Socialist movement in this country. It was in 1849 that Kingsley wrote his "Alton Locke," and it was at this time that, as "Parson Lot," he was most active. It was also in the same year that the Society for Promoting Working Men's Associations (of which the Christian Social Union founded in 1889 may be regarded as a spiritual descendant to-day) was formed at the instigation of Mr. Ludlow, with the inspiration of Maurice, and helped, among others, by Kingsley himself, by Hughes and Vansittart Neale.

The Society failed as an instrument of practical propaganda, for the numerous associations that it started rested on too insecure, and, perhaps, on too artificial a basis. They lacked growth from within, and in their form were in advance of their times. The promoters, many of them doubtlessly lamenting the rationalistic teaching of Owen, saw the industrial evils around them, they were appalled by the "sweating," and to them, as to many others, although they were hopeful of the possible future of the stores, the lack of capital seemed the great economic obstacle to reform, and the abuses to which men, as employees, were subjected, the great social danger. Higher aims were kept in view, and to Maurice above all others, dreamer of a divine order and profoundly moved by his conception of a present Kingdom of Christ, the economic proposals must have seemed but weapons of an inferior order. But he, like the rest, wielded them, and to provide capital, so that intermediate profits might be saved, and "labour" secure its fair reward, seemed to be the dominant practical teaching of the sad experiences of the time. But the event proved, as always, that it is as important to know how to use capital as to obtain it, and that, without either the power of management or the willingness to obey those who have, it is useless.

The hopes, the disappointments, and not less interesting, the mental attitude of the Christian Socialists of the time are reflected in the following extract from a preface to an edition of "Alton Locke," addressed to the working men of Great Britain, written by Kingsley, in 1854.

"Since I wrote this book five years ago . . . much has given me hope; especially in the

North of England. I believe that there at least exists a mass of prudence, self-control, genial and sturdy manhood, which will be England's reserve force for generations yet to come. The last five years, moreover, have certainly been years of progress for the good cause. . . . There is no doubt that the classes possessing property have been facing, since 1848, all social questions with an average of honesty, earnestness, and good feeling, which has no parallel since the days of the Tudors. . . . The love of justice and mercy towards the handicraftsman is spreading rapidly as it never did before in any nation upon earth; and if any man still represents the holders of property, as a class, as the enemies of those they employ, desiring their slavery and their ignorance, I believe that he is a liar and a child of the devil, and that he is at his father's old work, slandering and dividing man and man."

But however helpful the practical sympathies of others might be, Kingsley still saw that it was mainly by self-reliant action that working men must hope to advance.

"Your destiny is still in your own hands. For the last seven years you have let it slip through your fingers . . . How little have the working men done to carry out the idea of association in which, in 1848–1849, they were all willing to confess their salvation lay."

In the south, not only had the "self-governing workshops" failed, but no form of co-operative enterprise had been able to take root. Kingsley thus looks to the north and to Scotland, where, "thank God, the case has been very different." But it is the Rochdale system that has been advancing there, and it is to this system that, in 1854, the writer turns with most immediate hope—

"God grant that the workmen of the south of England may bestir themselves ere it be too late, and discover that the only defence against want is self-restraint; the only defence against slavery, obedience to rule; and that instead of giving themselves up bound hand and foot by their own fancy for a 'freedom' which is but selfish and conceited licence, to the brute accidents of the competitive system, they may begin to organize among themselves associations for buying and selling the necessaries of life. . . ."

For what was still regarded as a higher form of co-operation men seemed to have proved themselves to be unready, and with the sense of the underlying need for education, Christian Socialists, led by Maurice, were thus led to start the Working Men's College in London. Legislative advance also absorbed much of the thought of those among them who had legal knowledge, and the Industrial and Provident Societies Acts of 1852 and 1862, forming between them, to quote the words of Messrs. Ludlow and Lloyd Jones, "the Magna Charta of co-operative trade and industry," were directly instigated by this body of reformers, and notably by Mr. Ludlow himself. Industrial reform itself, however, became a hope for the future, rather than a practical propaganda for the moment. But faith in old ideas was not abandoned. "As for any schemes of Maurice's or mine," Kingsley wrote, in 1857, "it is a slight matter whether they have failed or not. . . . I believe that the failure of a hundred schemes would not alter my conviction that they are attempts in the right direction." The associations failed because the working men were "not fit for them." Kingsley doubts not, however, that "association for production will be the next form of industrial development," but (this from a letter in 1856) "it would take two generations of previous training both in morality and in *drill* to make the workmen capable of it." *

With modifications taught by experience, the co-operative ideas of these early Christian Socialists may still in the future be destined to prove themselves among the most powerful solvents of the industrial problems that the nation possesses, and at any rate the note of individual responsibility to which they give untiring emphasis is of too essential an economic importance for their propaganda to be ever lightly written down as "failure." For the moment, however, the proved greater difficulty of co-operation organized "in the interests of the producer" left the field comparatively free for purely distributive enterprise. With the partial exception of the corn-mills, which have always been annexes either of individual or federated stores, and, as Mr. Benjamin Jones, the historian of "Co-operative Production," points out, have "always been approached co-operatively from the consumers' side," it

* "Life and Letters," vol. i. chap. 14.

was thus co-operative distribution that characterized co-operative expansion in the immediately succeeding years, and as has been said, forms even to the present time the most distinctive development of the movement in this country.

In 1862, the year in which Parliamentary returns were first made by the Registrar of Friendly and Industrial Societies, legislation for the first time made it possible for one industrial society to hold shares in another, and the formation in the following year, of what is known to-day as the Co-operative Wholesale Society, has been, from the point of view of the movement, the most important sequel to this permissive Act. It would be difficult to exaggerate the importance of this great society, not only in carrying out the ordinary task of wholesale distribution, and in contributing to the financial stability of the stores, but also as having furnished the focussing point of some of the vexed questions of co-operative policy.

In 1868 the Scottish Wholesale Society was founded, and, in some of its branches in active partnership with the English Society, now holds, relatively to Scottish co-operative enterprise, a position equally important to that of its English forerunner in the south.

The by-paths of co-operation, as will be seen later, are perhaps destined to be as important as the main roads, and in this connection the amendment of the law relating to Partnership in 1865 is noteworthy. Up to this time, an agreement to share profits had been regarded as a contract of full partnership, and, in consequence, any workman admitted to a profit-sharing partnership, was liable to all the debts of his employer. In practice, this presented an almost fatal barrier to any legalized experiment for consolidating the interests of employers and employed, by making the remuneration of the latter not only, as they must always in the long run, depend upon, but vary with the profits of the former. The Act of 1865 did away with this obstacle, and led to the formation of industrial partnerships, some of which have figured prominently in the subsequent history of profit-sharing schemes, and which, although they failed to realize for long the great hopes that were at one time based upon them, are nevertheless valuable early examples of what may be done in this direction. Like the efforts of the Christian Socialists, they may be found to possess some of the exceptional significance characteristic of every industrial relationship that, in itself desirable, is introduced before men are quite ready for it.

In 1869 the first of the National Co-operative Congresses was held, and, proving the first of an unbroken annual series, led speedily to the formation of the Co-operative Union, a body that now represents the chief propagandist organization of the movement, as the Congress, which is now annually organized by the Union, does its chief deliberative assembly.

The Union is now closely associated with almost the whole movement, and the affiliated societies, according to the geographical position, are grouped into various sections. The Congress, with which supreme constitutional power rests, forms with what is known as the Central Board the governing body of the Union. The Central Board consists of representatives elected by the affiliated societies situated in each of the eight sections into which the kingdom is now again divided, Ireland having been reconstituted a separate section by a resolution of Congress in 1904. The Central Board meets twice a year, an executive known as the United Board meeting with greater frequency. The various sections are again subdivided into small "districts," mainly for purposes of conference and the discussion of matters of local interest. The whole scheme of the Union—District, Section, United Board (or Executive), Central Board, and Congress, with the various standing committees and special committees that are formed from time to time, together with the democratic basis of election that is adopted for the various bodies, provides a machinery that enables almost the whole Co-operative movement to realize its solidarity, and, be it for propaganda, for education, for expert advice, for legal protection, or for any other purpose, to ensure the expression of a thoroughly representative opinion. In 1906 the Union had 1230 societies in membership with it, with a total membership of 2,115,995.

The Co-operative Union, although the chief

central organization in the movement, does not stand alone even for purposes of propaganda, and certain other bodies have been formed because of what has been felt to be the limitations and the comparative inelasticity of the Union methods. In practice, this body is almost identified with the preponderating interests of the distributive stores and of the Wholesale Societies, and of co-operative production in as far as it is associated with these. But, since the days of Owen, the hope of actively extending the co-operative principle into wider fields of industry has never died out. Many have felt that the claims of "labour" as such must be conceded a special place in every entirely satisfactory development of the movement, and in 1884 the "Labour Association for promoting production based on co-partnership of the workers" was started, among the most prominent of its members being the late Mr. Vansittart Neale, the veteran Mr. Holyoake, and Mr. E. O. Greening.

Closely allied to this Association, although in actual formation preceding it by a year or two, is the Co-operative Productive Federation, founded primarily "to aid productive societies by combined action," and "to open up a market for the sale of their goods." It represents on a small scale the federal principle as applied to fields of co-operative production which are outside the Wholesale Society and the stores, and no society is admitted which does not practise profit-sharing with its employees. If these productive societies were at any time to increase in number rapidly, one of the chief functions of the Federation in its endeavours "to provide a market" would be to avoid overlapping, and—a danger which the critics of the detached societies never tire in pointing out—of avoiding also harmful forms of competition between the societies themselves. As yet, however, it is the day of small things with the Federation, the total trade of the affiliated societies at the end of 1904 amounting to less than £700,000.

The International Co-operative Alliance, founded to spread the knowledge of the co-operative movement, to promote international trading relations between co-operators of different countries "for their mutual advantage," and "to elucidate by international discussion and correspondence the nature of true co-operative principles," was established in 1895. It holds a Congress every two years, and although it cannot effect much to ensure the solidarity of a movement which in truth embraces such very diverse elements that a large measure of independent growth, even in the borders of a single country, is for a long time to come not only inevitable, but in the interests of experiment and freshness desirable, yet it marks a fresh step in the general recognition of the importance of co-operation as an element of industrial life, and of its increasing fame.

Coming back to the borders of our own country, we find three societies—the Agricultural Organization Society of Ireland, founded in 1884, the corresponding Society for Great Britain, founded in 1901,* and the Co-partnership Tenants Housing Council, founded in 1905—all of which give colour to the opinion that new work may lie before co-operative enterprise in the future, that will prove as important as that of the past has been.

* In 1905 the work of this Society in Scotland was taken over by a new organization, "The Scottish Agricultural Organization Society, Ltd."

With the mention of the Women's Co-operative Guild and the corresponding Scottish Guild, the list of societies that make for wider organization and propaganda ends, and ends not unfitly, by drawing attention to the great part, still often overlooked, that women can play in ensuring the practical success of the co-operative movement.

If we resolve in imagination the great system of co-operative distribution into its units of membership, we are at once taken back from the conference hall, the committee-room and the store to the home, with the woman on the threshold as the chief spender of its resources and oftentimes the chief influence. But as yet her eyes are often veiled; her power as a social force making for the health, comfort, and pleasantness of life is not seen, either by herself or by others. It is, however, upon the diffused power and interest of women that the vitality of a great part of the co-operative movement to-day in England in a great measure depends.

We have travelled far from the troubled early decades of the century, when the co-operative movement was in its struggling, experimental stage, . . . hopeful, but spasmodic and unassured. At the present time co-operative industry takes a recognized place in the national life. It is confronted by no legal disabilities, and is watched with sympathy both by the legislature and the public, apart, perhaps, from those with whose trading interests it conflicts, or those who are impatient of what seem to them to be its too slow methods of reform. Co-operators have, in fact, a fair field and a good deal of favour, and the spread of their movement will be determined by the alertness and the receptiveness of those to whom especially their appeal is made.

Co-operative industry still occupies but a small fraction of the great field of the national industry, and those who dream of the expansion of the present movement into the "Co-operative Commonwealth" of the future are probably hoping for a greater development than that to which its genius can ever lead. On the other hand, the movement is full of vitality and steady growth, and in some among the many forms it takes there is reason to think that it may be but on the threshold of its achievements.

PART I—THE STORE

CHAPTER I

SOME GENERAL PRINCIPLES

Essentially voluntary in character—Encouragement of Self-reliance and Freedom—Influence on Economic, Ethical, and Social Relationships—Its complex character.

THE forms which the co-operative movement takes are the outcome of three main forces: the pressure of industrial life, leading to attempts to improve the conditions of employment, and to secure better value for money earned; the exacting conditions of trade and manufacture making it clear, sometimes by success and not infrequently by failure, that real betterment can only be secured by the adoption of sound business methods; and, lastly, the spiritual energy, varying in strength, but running more or less through all the responsible leadership of the movement, with its reminder that co-operation will miss its highest mark if the character of its adherents be not raised, and if the heart of man be not touched.

Co-operation is, and has been characterized by a great self-reliance, and as regards its individual members its basis is perhaps more essentially voluntary than that of any other organized industrial movement. "Freedom," we know, is a question of degree, and organized social life is for ever nibbling at its edges; but in the co-operative life the limits of individual freedom are apt to be marked out more liberally than elsewhere. In this fact has indeed lain one of its practical difficulties, since co-operators have often not known how to yield to guidance in management, and submission to accepted leadership is one of the lessons that they have often still to learn. The fact remains, however, that in spite of many practical failures to give it satisfactory expression, the principle of voluntaryism has underlain in an exceptional degree all healthy co-operative development. The element of coercion hardly enters either for the individual co-operator in connection with his fellows, or as regards his relations to those who remain outside the co-operative body, and in this fact lie elements of ethical superiority alike over militant trade unionism which, when strong enough, practically makes membership compulsory, or over any form of political propaganda that aims at power of some kind at the expense of others.

Alike in theory and in practice, co-operation aims at leaving unweakened the industrial and social freedom of the individual life. Every unit of the movement in taking the step that brings him within the ranks of the co-operators thus takes it as a free agent. In the vast majority of cases he takes it, it is true, because he considers that the step will bring him some personal advantage; but when this self-regarding freedom of action is combined with, or, as so often happens, leads to the belief that the co-operative principle is worthy of propaganda for the sake of others, we are then confronted with emanations of the co-operative faith, and of the spiritual impulse of a voluntary comradeship that is trying to give effect to the co-operative motto that bids each man take thought for the welfare of all. This is the "gleam" that the best leaders of the past, and many in the present, would fain have seen, and see the rank and file follow, and the greatest failures of the past, and the greatest dangers of the present, have been, and are, connected with blindness to this light, which to-day prosperity, and the maintenance of prosperity, no less than difficulty, sometimes overclouds.

The most abiding interest of the movement, however, is due to the demand that it is still apt to make upon all who join it that some effort shall be made, so that all men may be able to look more courageously and more self-reliantly towards the future. For the movement is a hopeful and progressive one, and as demonstrating a great reserve of social force, capacity, and good will, it is perhaps able to establish a firmer hold upon the public imagination than by its power to make for a more widely spread security and comfort, by its merits as an industrial system, or by its power to increase the economic usefulness of the individual life.

In all of these ways, however, industrial co-operation is important. In spite of its manifest limitations, alike in aim and in achievement, it is able to exercise a distinctive influence in such very different fields as the conditions of economic and social well-being, the ethics of industrial relationships, and the national welfare.

It is, however, in itself, a very complex affair. "Co-operation" is always "more" or "less" completely co-operative, and the examination of the various forms it takes makes it clear that one or other of its manifold applications exercises also some special form of influence—social, ethical, or economic. Such examination would, however, show but ill-defined boundaries, for the forms of influence mentioned are themselves overlapping forces in life, often inextricably interwoven. Although no such analysis as that suggested is possible, therefore, it may be noted that co-operative distribution, undertaken as it is "in the interests of the consumer," bears most directly upon the questions of material and domestic well-being; the applications of co-operation to productive enterprise bear most directly upon the questions of industrial ethics and industrial relationships, while the application of co-operation to agriculture and the various forms of profit-sharing in what is for the most part capitalistic enterprise, gives rise to the greatest economic questions that affect the national welfare. It is to the consideration of the first of these that the following section is devoted.

CHAPTER II

THE DISTRIBUTIVE STORE

Conditions of Membership—Shares—Fixed Rate of Interest—Supply of Capital—The Committees of Management—Staff—Prices and Profits—Dividend—Credit—Competition—High Dividends—Non-co-operative Sources of Supply—Their Increasing Efficiency—Differences in Size of Societies—Inception—Examples of Growth—London—The Number of Societies—Recent Growth—The Danger of Overlapping—" Loyalty "—Gain and Loss—Purchases per Member—Examples—The Influence of Local Conditions—Average Earnings—The Margin of Non-co-operative Expenditure—Table—Effects of Wages and Price Movements—Expansion and its Limitations—Comparative Table

THE Co-operative Stores, in the form that may be said to have started some sixty years ago, now represent, so far as Great Britain is concerned, the most flourishing side of the co-operative movement, and some of their leading features may now be briefly described.

Membership is open practically to every one of either sex over sixteen years of age. The constitution is thoroughly democratic, and the most exacting qualification required to give voting power for the executive committees, by which the societies are governed, is that of holding a fully paid single one-pound share.

This share may be paid for by easy instalments, but a common practice is to pay a shilling on application, and to leave the dividends on subsequent purchases to accumulate, no further calls being made.

By the Industrial and Provident Societies Acts, under which almost all the societies are now registered, no member can hold shares for a nominal value of more than £200; but when this sum is reached, the society, if it can use the money, can borrow it as loan capital. The supply of sufficient capital is no longer a difficulty, and in the case of some societies the maximum of share capital per member is voluntarily fixed at a smaller amount than the legal limit of £200 in order to keep the supply within manageable and safe bounds.

No member has more than a single vote, however many shares he may hold, and with the occasional exceptions of the first, the shares are withdrawable but not transferable. This arrangement is the simplest, and appeals most to the great mass of those whom the stores attract and desire to attract; but in the case of any special adverse circumstances leading to something of the nature of a panic, in the case of any widespread local depression in trade and employment, or if any considerable section of the members become for some reason disaffected and leave the society, it is evident that this power of easy withdrawal may at any time become a source of danger. The provision of adequate reserve funds, and proposals to give greater power to the committees to suspend the right of withdrawal "whenever circumstances arise that make this step imperative in the interests of all the members," were among the subjects most seriously discussed at the Congress of 1904, and a resolution was passed then advising the societies to take the matter into their serious consideration and to secure any necessary amendment of their rules.

Interest, always fixed, is generally paid at the annual rate of 5 per cent., a rate that is somewhat abnormally high in these days, and the tendency is to reduce it.

By the automatic accumulation of dividends, added to the attraction of a substantial rate of interest, the capital of the stores often outgrows needs, and the utilization of these surplus resources is a practical problem that has found its chief solution in investment in one of the great Wholesale Societies, and, to a still greater extent, in house property—the houses thus acquired, sometimes having been also built, and in many cases let to members of the society.* Another important outlet for the use of capital is found in the various branches of "productive" business sometimes undertaken by the stores.

Subject to the general supervision of the shareholders' meetings, generally held half-yearly, management is vested in the committee, and its officers are elected by the shareholders, and, as regards the actual conduct of the business of the store, in the manager, who is himself appointed by the committee, and thus by a process of indirect election. Subordinate employees are often appointed on the recommendation of the manager, and these, together with the manager himself, generally stand in the same relation to the committee as does the corresponding staff of ordinary joint stock companies or private employers.

* See ch. xix.

The prices at which goods are sold are generally fixed at the general level of the district in which the stores are situated, and no attempts are made, as by such joint stock institutions, as the Civil Service, or the Army and Navy Stores, to attract members or custom by exceptionally low prices. The economic attraction of the working-class store is not found simply in a series of advantageous purchases, but in the knowledge that the profits accruing on the series, after allowing for all administrative and working expenses, will be at the disposal of the purchasing member in the shape of dividend.

It is in this way that the element of retail profit, and, so far as goods sold have been purchased from a co-operative Wholesale Society of wholesale profit too, is eliminated from the price basis of the store, and that the purchasing member may not incorrectly be described as ultimately getting his goods at cost price.

Purchases are recorded by giving a metal check or some other token, and upon these, when presented at the end of the quarter or half-year, representing as they do the total of the purchases made by the individual member, the share of the dividend that has been declared is paid, or, if not withdrawn in cash, is credited to him as additional share capital. Dealing in dividend-tokens between members is always deprecated, but, at a sacrifice, they can generally be presented at the office of the society and realized upon before the end of the half-year.* To non-members who deal at the stores, half the amount of the full members' dividend is generally paid.

The impression is widely prevalent that the "ready-money" principle is universally adopted at the stores, but this is erroneous, credit being given by a large majority of them—according to the figures presented at the Congress of 1904, by 1401 societies in the United Kingdom as against only 507 which worked exclusively on the cash principle. The practice is deprecated,† and the principle of cash payment is

* The following is a "Caution" printed on the reports and balance-sheets of one large society in the north of England:—"Check buying: Members are warned that the purchasing of checks is a violation of the Rules of the Society, and renders the purchaser liable to an action at law, expulsion from the Society, and the forfeiture of all benefits. Checks brought in which have been purchased from any other person are not entitled to rank as claims for dividends, and all such cases coming under notice of the directors will be severely dealt with."

† As, for instance, by Mr. J. C. Gray, in his interesting

extolled by co-operators, but it is, nevertheless, unlikely to be made in any sense compulsory, even in the sense of not allowing the credit given to exceed the accumulated savings of the member. Not only is the practice very difficult to check, and much more so to abandon, when it has been once adopted, but many arguments are advanced in its favour. It is said, for instance, that a binding rule forbidding

analysis of the subject: See "The Credit System as practised by Co-operative Societies," published by the Co-operative Union in 1888. Mr. Gray, now the secretary of the Union, is still one of the severest critics of the practice. The question is referred to as follows, in the report of the Central Board :—

"The growth of credit trading in the movement is such as to cause serious anxiety to those who have its best interests at heart. We have on many occasions called attention to this growing evil, and the practice has been condemned over and over again by Congress. Unfortunately, however, as in many other matters, the spirit of the movement, and the ideas of Congress, do not always actuate and guide societies in the conduct of their business, and many of them, anxious to maintain their trade and extend their business, follow on the lines of their competitors, and sacrifice principle for gain. During the year many societies have altered their rules so as to recognize and legalize credit trading to an extent they had never done previously, and a glance at the annual returns published by the Registrar shows that it is now officially recognized and carried on by the majority of the societies."

A committee representative of the United Board and of the Women's Co-operative Guild has been appointed "to consider the best means of dealing with the matter, so as to induce societies, as far as possible, to adopt a system of cash trading." [*Congress Report*, 1906, pp. 138–9.]

all credit would often involve unnecessary hardship, or drive the member away from the society altogether, while the special circumstances that may arise in the case of a labour dispute, of bad trade, or of irregular employment are urged, and it appears that all that can be hoped for is that credit will still be regarded as the exceptional method of buying to be claimed and sanctioned only when circumstances appear to justify departure from what is still regarded by the vast majority of co-operators, at least in theory, as the golden rule of cash payment. The increasing extent to which this rule is departed from is, however, a very serious matter, and one that, if not carefully watched and checked, may, in a few years, do much to imperil not only the financial stability of many of the societies, but, which is far more important, their power to safeguard the home by safeguarding the purse.*

* In the cases in which, though forbidden by rule, credit is nevertheless given by registered societies—and nearly all are registered nowadays—it would appear that the office of the Chief Registrar of Friendly Societies could with great advantage make itself more effectively felt than at present, if only, as a first step, in the sense of asking for an explanation of the irregularity. It is to be noted, however, that official conformity with the rules in this matter may still be accompanied by their

The vast majority of co-operators belong to societies paying over 1*s.* 6*d.* and less than 3*s.* 6*d.* in the pound as dividend. Out of 1328 societies comprised in the Board of Trade returns * as to the dividends for 1905, only 2·2 per cent. of the total membership of over two millions belonged to societies paying 1*s.* in the pound or less, and only 9·3 to societies paying over 3*s.* 6*d.* in the pound. The average rate of dividend paid was 2*s.* 6¾*d.* in the pound—a figure from which there is little variation year by year, although the actual figure for 1905, it may be noted, shows a reduction of 1*d.* in the pound as compared with the years 1899–1902. This decrease is doubtless due to recent agitation in favour of lower prices and dividends, with the view of bringing in a poorer class of consumer.

The amount of dividend paid by individual societies varies roughly according to the margin

practical non-observance. Not infrequently this takes place through branch managers, who, through kindness, or weakness, or because they think that it answers their purpose, give credit on their own responsibility. In some cases they do this genuinely at their own risk; in others the practice is "winked at" at the headquarters of the society. The regularizing of the whole question therefore bristles with difficulties.

* See *Eleventh Abstract of Labour Statistics.*

of profit that the competition of the outside market makes possible, and the amount is thus generally lower in great centres of population where the consumer is catered for with every imaginable device and attraction — poster, advertisement, "leading lines," bonus gifts on purchases, payment by instalments, or low ready-money prices, as the case may be—offered by the ordinary retailer, than in smaller places where competition is less keen, and where the individual rate of profit generally rules higher. In some small centres of population, most frequently in the North of England, the co-operative society may be so powerful as to be almost independent of outside competition, and it is under such circumstances that the highest dividends are most likely to be paid.

Great centres of wholesale exchange, of which London is the pre-eminent example, tend to keep down the level of retail prices and profits—and thus of dividends, and this affords one of the reasons that adds to the practical difficulty of the co-operative movement in the Metropolis. It is, however, only one, and the normal disintegration of the population there, combined with the difficulty of finding or creating the social or industrial bond that makes men know each other, and thus provides the personal basis upon which the movement so largely rests, is perhaps a still greater obstacle. Co-operative success, moreover, always presupposes a certain amount of co-operative knowledge and of familiarity with its constitution and objects, and this knowledge or familiarity the vast majority of the people of London are entirely without.

The importance of a co-operative tradition and of local solidarity as aids, and of outside competition as a hindrance, are probably reflected in the differences in the average rates of dividends that are earned by the societies in the various geographical sections of the Co-operative Union. Thus, in the Southern, the average is 1*s.* 3¼*d.*, in the South Western 1*s.* 5*d.*, in the Western 1*s.* 5½*d.*, in the Midland 1*s.* 8½*d.* in the North Western 2*s.* 9*d.*, and in the Northern 2*s.* 11*d.* It may be noted that in the case of individual societies the dividend not infrequently reaches a considerably higher figure than the highest average just mentioned, but the exceptional rates sometimes earned are generally a sign that the management is somewhat indifferent to the circumstances of the poorer members of the community, who cannot be expected to pay exceptionally high prices however large the prospective return in the shape of dividend may be. On these grounds, therefore, very high dividends are often rightly deprecated, and are interpreted rather as signs of inconsiderate than, as might have been assumed, of exceptionally capable management.

In spite of the importance of the attendant social features of the co-operative stores, their success has, after all, been attained in the first instance because they were able to discover the economic weakness of many of those whose functions they usurped. The competitive rival against whom the store finds it hardest to make headway, is the retail distributor who has a well-equipped, well-managed, reasonably priced shop, in which a "living" is being made on a capital and a turnover that might reasonably yield one. It is the small shops which can only buy in small quantities and thus at a disadvantage, the turnover of which is small and which therefore are obliged, if they are to succeed at all, to make large profits on whatever business is done, that are most easily driven from the field by the stores. These shops are, it is true, often strong in their power of ready adaptability to the needs of the body of near neighbours who generally provide them with their customers, but, in general, they stand for inefficiency in the *rôle* they are supposed to play and for economic waste.

This inefficiency, combined with much waste of time and energy, is still characteristic of much of the business of retail distribution, but rapid changes are taking place, and it is probable that in the aggregate the ordinary retail sources of supply have never been so efficient as they are at the present time.

In this connection it is worthy of special note that people of every class, except the very poorest, are becoming less and less dependent on local dealers, save for very perishable or bulky commodities, such as meat, bread, and most vegetables. The parcel post, press advertisements, printed catalogues and circulars, the local agencies of big enterprises, and the general increase in the facilities for communication and transport, combined with the greater knowledge to which all this leads of the prices that are ruling in other parts of the country, are making it increasingly difficult for even the remoter centres to maintain an inflated local level of prices, while in greater centres, save for articles in connection with which there is some special limitation of supply, such as the need of exceptional skill or taste, the margin of profit tends to be smaller and more uniform. To a certain extent this change has probably been hastened by the stores themselves, but for the most part it is due to wider competitive influences, analogous to those that, apart from artificial fiscal checks, are consolidating the markets and levelling the prices of the civilized world.

There are other reasons, such as the administration of the Weights and Measures Act, and the Food and Drugs Act, that go to show that, merely as parts of the machinery of distribution, the *rôle* played by the co-operative stores is probably becoming less and less important, especially as regards the comparatively substantial class for which they for the most part cater, and by which they are at the present time chiefly used. For the classes below it is otherwise, and to this point we propose to return.

It would be very misleading, however, to regard the distributive store as being merely an ingeniously devised centre of retail trade, that made it possible for the working-class consumer to buy advantageously through the medium of the dividend, and, if he be so minded, to save, almost automatically, by the same powerful instrument.

To some among their own members, and, it is often thought, to an increasing number, these two functions constitute the sole attraction, and it is undoubtedly to them that the stores owe the greater part of the economic strength upon which all else must be built up.

But the store stands, or ought to stand, for much more than for successful trading and for thrift. In the supply of commodities that are not produced under bad conditions, in the payment of fair wages, in its implied and practical protest against the grosser inequalities in the distribution of wealth, and, finally, in its demonstration of the spirit of fellowship, which is the higher quality and manifestation of industrial association, it ought always to find a yet higher justification.

The store provides, in addition, a training school that teaches working men how to administer capital, gives them fresh insight into its fleeting and destructible character, and that new sense of responsibility that the control and use of capital brings. The store is not infrequently a centre of education, and from it also, although in ordinary party politics the societies are neutral, there often emanates a public spirit that tells effectively and wholesomely upon local administration. The social value, in short, of a store that is co-operative in spirit as well as in form, is as great as, if not greater than, that which is purely economic. In view, therefore, of the influences that have been mentioned, that tend to diminish the utility of the stores to consumers of the well-to-do working-class as mere centres of distribution, and in that respect to weaken their position, it will be increasingly necessary that due prominence be given by the leaders of the movement to what may be described as their higher uses.

The societies to which the foregoing pages have referred are of every size, ranging from a tiny affair with four members and aggregate sales of less than £150 in the year to the colossal enterprise centering at Leeds with nearly 50,000 members and annual sales of more than a million and a half.

But, however large any of these societies may be to-day, they have nearly always started on a small scale.

> " Some working-man of somewhat more intelligence, or with somewhat more influence than his fellows, hears of what is being done by working-men co-operators in England and Scotland. He talks the matter over with his friends and comrades, obtains and circulates tracts, and finally they make up their minds to see if they can start a society." *

In this, or in some similar personal way, the impulse is nearly always given, often helped nowadays, perhaps even initiated, by the action of the agent of the Joint Propagandist Committee of the Co-operative Union and the Co-operative Wholesale Society, a body that is always on the look-out for

* Acland and Jones, " Working-men Co-operators," p. 33.

chances of expanding and strengthening the movement. The provisional committee is somehow formed, the capital collected, and probably advised and provided with model rules by the Co-operative Union, the beginning is made.

Conditions as regards capital, the supply of which, at the outset, is always limited; the need of experience in management, and the normal timidity of those who feel that they are risking something of their own, and know that they will have to make their way in somewhat unknown fields, make these early days nearly always also the days of small things, even if the initial difficulty of securing the adherence of considerable numbers has been overcome.

The Rochdale Pioneers furnish the historic instance of expansion, because the circumstances of their humble origin are so well known, and because the part that their constitution and success have played in the subsequent development of the movement, has met with such universal recognition. The growth of their society, however, from the twenty-eight members of 1844 to the 12,800 members of to-day, a figure that represents, from various local causes, some set-back from the total reached a few years ago, is by no means exceptional. The Royal Arsenal Society, at Woolwich, for instance, now (with the exception of Plymouth) the largest society in the south of England, and one of the largest in the country, began in 1868 on even a smaller scale, with twenty members and a trading capital of £7. But during 1905 the sales of this society amounted to more than half a million. More than £200,000 had been invested in house property, most of it in dwellings erected by the society on its own estates; the share capital amounted to £347,000, and the net profit, after allowing for interest on share capital to the amount of £13,694, was £55,308, yielding a dividend of 1*s.* 6*d.* in the pound, and £2988 for the 1134 employees as their special extra share of the earnings, while something over £1000 was spent for educational purposes. The membership was 26,146.

Figures such as these, however, while illustrating the possibilities of expansion, are realized only when the circumstances are favourable, and, as already mentioned, London as a whole is as yet almost untouched by co-operation. The place is, perhaps, almost too vast for the success of a movement which depends to so great an extent upon men knowing who their neighbours are. The residential working-class districts are generally remote from the centres of employment. These centres are often on a small scale, and even if they are not, the occupiers in any given district are generally detached units dwelling for the most part in the midst of a sea of unknown faces, for, both socially and industrially, the bond of neighbourliness in London is notoriously weak. Finally, in no place does the force of retail competition make itself so effectively felt as in the Metropolis—the cheapest as well as the dearest place in the world in which to live.

It is worthy of note in connection with the rate of growth that in England and Wales, while from 1880 to 1894, 601 retail distributive societies were started, and 366 ceased to exist, leaving a net gain in each quinqennium of 78 societies; from 1895 to 1899 only 127 started while 121 ceased to exist, leaving a net gain of only 6 for the five years.

The same feature is shown by the corresponding returns for Scotland. Here 110 societies were started from 1880 to 1894, and 54 ceased to exist; while from 1895 to 1899, 18 started and 20 ceased to exist, showing a net loss of two.

During the years 1902–1903, however, the returns of the Co-operative Union again showed a considerable aggregate increase, and in 1903 the maximum was reached of 1481 societies making returns.* In 1905 the corresponding number was 1457.† It is probable that the great majority of societies recently formed are established in well-chosen districts, since it is the policy of the Co-operative Union and the Wholesale Society, through their joint propaganda committee, to aim at consolidation and amalgamation, and, as far as they can control matters, to limit the formation of new societies to those districts in which there is no well-established society within a reasonable distance, where, therefore, there is no chance of getting a branch established, and

* *Congress Report*, 1904, p. 542. † *Ibid.*, 1906, p. 550.

where, in consequence, if any co-operative start is to be made, a new society must be formed.*

Thus, recent years seem to reflect not so much the spread of the movement as the consolidation of the different societies, in itself a matter of satisfaction, and especially so in as far as it indicates, perhaps, some abatement of the difficulty that co-operators, hardly less than others, have often had, and still have, to contend with, namely the competition of a harmful kind among their own societies that had been placed in, or that had grown into too close a proximity to each other.

The total strength of the co-operative movement is not, it may be noted, measured simply by the number of societies and the total number of members in proportion to the total population, but by the extent to which the stores are used, or as co-operators themselves would say, by the extent to which members are "loyal" to them.

The most frequent charge made against the less thoroughgoing of the co-operative members is this one of "disloyalty," by which

* *Congress Report*, 1906, p. 115.

is meant the habit of buying elsewhere, goods that the store could have supplied. This diversion of custom is generally determined by a comparison of relative prices, although not infrequently it is simply the maintenance of a personal and neighbourly connection already formed, or it may be simply a liking for a greater variety than any single centre of supply, no matter how well stocked, can hope to offer.

If cheapness alone be considered, unless there is good reason to suspect that the alternative sources of supply deal in goods that are not only of lower price, but are produced under anti-social conditions, it is innocuous, perhaps it is even desirable, in the interests of efficient store management, that the consumer should thus, within reason, pick and choose, and this power of selection that every one possesses is freely used in practice.

The co-operator, however, soon learns that the cheapness of the individual transaction is by no means the complete measure of its advantages, and it is upon the solid personal benefit secured by a multitude of transactions that quarter by quarter are seen to bear the solid test of adding to the comfort and well-being of the home, that co-operative distribution mainly depends for its success. In this respect, however, the test is much the same as that of all well-ordered expenditure; but, as we have seen, the co-operative system has learnt how to combine with this sundry other advantages that facilitate saving, weaken the credit system, widen interests, and, especially for those selected by their fellow-members to fill the various offices of the society, create a sense of fresh and useful responsibilities, and often make for fellowship and friendliness among neighbours and fellow-workers.

All these things represent an aggregate advance, alike on economic and social grounds, and perhaps the only drawback is found in the harder position in which retailers, generally of the smaller kind, are placed in consequence of the competition of the store. Not infrequently these have to give way entirely, and cases of individual suffering may then result. This effect of competition is, however, analogous to that which follows from the extension of large private firms, the successful establishment of the branches of which tends to have almost exactly the same competitive effect as the prosperity of a co-operative store. Society rarely secures any economic gain without offering up some sacrifice, and if the gain is real there is nothing for it but the adaptation to the altered conditions, often by painful processes, of those whose position is assailed. A change of custom is, it is true, often determined by mean motives, and the desire for cheapness "at any cost"—even at the cost of the health and happiness of unseen workers, is responsible for some of the worst evils of industry. But no class can complain of "unfairness" if it should find its services no longer requisitioned; in a free industrial community there is no such thing as a prescriptive right to a weak economic position.

The loyalty of co-operators to their societies varies largely, not only, of course, as between member and member, but by areas. It is in the smaller, rather isolated centres of some mining or manufacturing industry that the highest average purchases per member are most often reached. Such figures, for instance, as those for the Murton Colliery Society in Durham, with more than 1200 members in

1905, and sales equivalent to purchases of about £55 per member in the year, rare anywhere, are never found in large towns, even in those situated in counties in which co-operation is strongest. Thus, in the same county of Durham, the average amount of purchase per member at Gateshead is about £33 per member, or about £1 per member more than that of such a great co-operative centre as Burnley in Lancashire. Some allowance has, doubtless, to be made for the higher average spending power of the Murton members, but the main explanation of the contrast which the figures offer is due to the greater alternative sources of purchase that towns always offer. In Woolwich, for instance, we find that in spite of a very active policy of branch extension, leading sometimes to their establishment almost in advance of the rapidly increasing population of the neighbourhood, the nearness of London and the growing number in recent years of well-appointed private shops, are influences that make themselves felt, with the result that the average amount of purchases per member was here a little less than £20 in the year. In Woolwich, the backbone of the Society is furnished by the large body of Arsenal employees, many of them men brought up amid North-country surroundings, and thus both by the conditions of their employment in big works, with the easy growth of public opinion and the adoption of uniform trading practices that such employment fosters, fitted by early experience to apply the co-operative principle even in a semi-London environment. In the same way in Stratford, just beyond the borders of the County of London, the great body of men employed at the Great Eastern Railway Works supplies a suitable and, it may be said, even necessary nucleus. In the latter place is thus found the only other large society in or near London, and this showed an average of purchases in the year of something over £24 per member, a figure that, when, as with Woolwich, the proximity of London, and the alternative sources of supply on the spot are taken into account, speaks well for the management of the society and for the hold it has upon its members. But the average is still about 57 per cent. lower than that of the little Durham mining village.

If we turn to such a populous centre as

that of Plymouth, although the society is a strong one, with nearly 35,000 members and sales amounting to £625,000, it is remote from the more concentrated energy of the co-operative movement, and the average annual purchases per member are found to drop to about £18. This figure may perhaps be contrasted with that of the great society which has Leeds for its head-quarters—a still greater centre of population, but situated in a northern manufacturing county in which co-operation is particularly strong. Here the purchases per member averaged nearly £32 in the year. For the whole of the United Kingdom, the total membership, according to the returns of the Central Board of the Co-operative Union, having been 2,153,185 and the total sales £61,086,991; the average was £28 7*s.* 0*d.*, a figure that is equivalent to about 10*s.* 11*d.* per week, or, deducting the amount returnable on the average in the shape of dividend, about 9*s.* 8*d.* per week.

As will be seen later, this average is approximately stationary, and it clearly proves that a large margin of the expenditure of the members of the various societies that might have been spent at their stores is diverted elsewhere.

It is impossible to state with precision what this margin may be, since we have no exact knowledge as to the average incomings and spending powers of co-operators. As of the working-classes at large, so of these, any figures that may be given can only be approximations to the truth, since the elements of the problem are so complex and so variable. In connection with the stores, one special point further complicates the problem of estimating spending power, since the individual household is not infrequently represented by more than one member of the family, although representation by a single member is still the general rule. Mr. Greening, in his presidential address to the Co-operative Congress of 1904, considered that it was a safe estimate to take the 2,116,127 members then returned for societies of every class, as representing 8 millions of the people; and if on an average, we calculate that one family in every five has more than a single representative in the stores, we shall probably not greatly err. On this basis it would be necessary to add one-fifth to the weekly average mentioned above in order to arrive at the amount spent per family, and the figure would thus be 11*s.* 7*d.* instead of 9*s.* 8*d.* per week.

If, including supplementary earnings, we take 34*s.* per week as the average incomings of the co-operative household, a figure that is probably below the mark, and the proportion spent on food at 55 per cent. of this total, the weekly average expenditure on this head alone has a potential weekly rise to about 18*s.* 6*d.* The average working-class expenditure on boots and clothing for the same class of family may perhaps be put down at 15 per cent. of incomings, or, on a total of 34*s.*, at about 5*s.* per week.* Certain other items of expenditure might be, and often are, made at the stores, such as furniture, hardware, coal, oil and tobacco, but without allowing anything for these, it appears that the weekly average spent at the stores might, if they had all the necessary departments, could make themselves sufficiently attractive, and if the members resisted the temptation to turn their feet towards other sources of supply, be increased by about 100 per cent.

* Cf. *Board of Trade Return*, Cd. 1761, 1903, pp. 209–217.

The following short table will show that there is little if any tendency for the stores to strengthen their position in this respect, and that while total sales and total membership are steadily increasing, personal loyalty, as measured by purchases, remains almost the same.

WORKMEN'S ASSOCIATIONS FOR RETAIL DISTRIBUTION IN GREAT BRITAIN

Year	Members	Sales	Sales per member to the nearest shilling *		
		£	£	s.	d.
1881	546,712	15,392,127	28	3	0
1891	1,043,634	30,568,069	29	5	0
1896	1,354,021	36,621,980	27	1	0
1897	1,462,858	40,074,960	27	8	0
1898	1,532,832	42,521,258	27	15	0
1899	1,610,320	44,985,490	27	18	0
1900	1,707,011	50,053,567	29	6	0
1901	1,789,658	52,670,719	29	8	0
1902	1,889,178	55,209,122	29	4	0
1903	1,982,510	57,373,252	28	19	0
1904	2,072,792	59,161,707	28	11	0
1905	2,146,242	60,901,553	28	7	0

The aggregate advance which these figures show must be interpreted so far as the money columns are concerned in the light of two facts; (1) of any changes that may have taken place

* No deductions have been made on account of dividend. It may be noted that taking the average dividend of 2*s*. 6*d*. in the pound, the total dividend accruing will average £3 10*s*. 11*d*. per member, or something like 1*s*. 4*d*. per week.

contemporaneously in the level of wages, and (2) similar changes in prices. It is, therefore, especially noteworthy that from 1881 to 1896, the movement of prices was, with slight exceptions and interruptions, steadily in favour of the working-class consumer.

About four-fifths of the total business transacted by the English Co-operative Wholesale Society is in food-stuffs of various kinds, a proportion that probably only slightly exceeds that of the transactions of the various stores since food always forms by far the most important group of articles in which they deal. For instance, two recent half-yearly balance-sheets which are at the writer's side, of a rather small society with less than a thousand members, show for the twelve months under five heads of incomings the following amounts: boots, £1,306; tailoring, £333; drapery, £2,037; coals, £1,276; while the fifth, "grocery and bakery," is responsible for no less that £11,291. In this society there is no butchery department, neither is milk supplied. In this case, the sales of food thus formed about 69 per cent. of the total, but a corresponding analysis of the half-yearly accounts of four large societies, in which, with the exception of milk, almost every possible branch of co-operative food supply was always undertaken, shows the following percentages: 75, 77, 78 and 85. The aggregate retail sales of the four societies in question amounted to £1,326,149 for the half-year, and out of this total £1,065,920 was spent in food, or almost exactly 80 per cent.

We shall, therefore, get a very fair, if not complete, indication of the way in which the movement in prices is likely to affect the money value of the transactions of the stores, if we know what the general trend of retail prices for food is in any given period. The table prepared by the Board of Trade* gives us valuable fresh information on this point, and indicates a fall in the average prices of food in working-class expenditure of no less than 49 per cent. between 1881 and 1896. The cost of boots and clothing is also given in the returns, and this too diminished, but much less rapidly, the fall amounting to just under 10 per cent. in the same year. In the cost of fuel and lighting, little change took place, but rent, on the other hand, increased as much as that of

* See *Return* [Cd. 1761, p. 224], Cd. 2337, [1904], p. 33.

boots and clothing fell. For furniture and hardware, no particulars are given, and they rarely figure largely in a working-class budget, but it is however well known that during the whole of the above period the general direction of prices in these commodities was markedly downwards.

Of the items above mentioned, fuel and lighting do not greatly affect co-operative spendings, and furniture and hardware might be omitted from our calculations, but the effect of the rise in rents must be taken into account, since it tends to check spending power in other directions. Allowing for this, but also taking into account the very large proportion of co-operative expenditure which goes in food, we shall not be far wrong in estimating the increase in the effective spending power of co-operators at 31 per cent. between the years mentioned. Thus, if the sales per member had been only about £19 9*s.* instead of £27 1*s.* in 1896, the level of 1881 would have been maintained.

Since 1896, however, the tendency has been for prices to rise, the change in the average retail price of food showing according to the table mentioned above, instead of any fall as in the earlier period, an advance of 14 per cent. between 1896 and 1901. Rent and clothing remain as comparatively fixed charges, but noting such differences as occurred, and again allowing for the preponderating influence of changes in food charges, it may be calculated, again leaving variations in wages out of consideration, that the effective spending power of co-operators declined about 11·5 per cent. between 1896 and 1901. The 20*s.* of 1881 instead of going as far as about 26*s.*, as it would have done in 1896, contracted to something less than 18*s.* Thus, while the apparent fall from £28 3*s.* to £27 represented a real and considerable advance when measured in quantities and not in money values, the apparent advance from the latter figure in 1896 to £29 8*s.* in 1901, represents in reality a slight decline, since about £30 3*s.*, instead of the actual figure reached, would have been necessary to maintain the real average of 1896.

In the later period also, much more markedly and continuously than in the former, the general conditions of employment and the trend of wages would also have led us to expect a higher average. From 1899 to 1901, no important field of employment was greatly disturbed by industrial disputes, as had been the case in the two preceding years; trade was buoyant and wages were exceptionally high. During 1902–1906, there has been some set-back alike in prices and wages, and some reflection of this is probably excusable in the somewhat smaller average outlay per member shown for these years.

The general conclusion, however, to which a study of the table on page 72 leads us, after all allowance is made for the fact already mentioned that not infrequently more than one member of the family joins, a practice that, as women are learning to take a greater interest in the stores, is tending to become more common when the rules of the societies make it possible—is that, while the movement is strong and steadily growing, the rate of growth, or at least of expansion, is somewhat slackening, and that the average trading hold upon individual members, reflecting, as the figures do, a somewhat limited loyalty, is either stationary or tending to become slightly weaker.

It may seem to some readers of these pages that contradictory arguments have been advanced with regard to the co-operative movement; since, on the one hand, influences have been mentioned that are tending to weaken the position of the stores, and on the other much has been written emphasizing their continued and increasing vitality.

It will be found, however, that there is no incompatibility between the two lines of argument. Enough has been written to show how great are the economic and social advantages of a well-managed society. Even in a district in which the economic gain as a shopping centre is not, or perhaps is no longer, great, when once a start has been successfully made, when considerable numbers have invested their savings and developed financial interest in the continued prosperity of the store, and when on wider grounds they have become interested in its management and its welfare, a body of opinion and of goodwill exists that goes far towards ensuring stability and even growth, in the face of strongly adverse competitive conditions. The store will be found to possess an economic and social momentum which it will need either great mismanagement or great disaster to destroy or arrest.

The fields, moreover, in which there is good ground for thinking that there are favourable openings for new societies, are far from being exhausted, while in the increase of population and the development of industry, fresh fields are being constantly created. Moreover, as the children of co-operators grow up and become in their turn heads of families and householders, many of them would almost, as a matter of course, associate themselves with the neighbouring society, and the movement is old enough now to be reaping the full benefits of traditional or inherited allegiance and experience. "I have seven sons," said a woman of nearly seventy years of age, at a recent co-operative congress, "all married and all co-operators."

Thus, leaving out of account the increasingly effective help that may be given by central organizations whether in the form of advice, as by the Co-operative Union, or by way of facilitating the actual business of store management, as with the Co-operative Wholesale Society, there are more than sufficient grounds to account for the continued development of the movement.

The table on page 81 gives the most salient particulars of this growth since 1881, and for the purposes of comparison, and in order to make it easier to keep in view the position of the co-operative store movement in relation to yet wider interests, figures of population, total exports and imports, and income-tax assessments have been added. For it seems well, since every industrial action helps or hinders, makes or mars, the national life, and since industrial "movements" are never self-contained, but always dependent for their vitality upon that of the larger life in the midst of which they flourish, that we should be reminded that even the most notable of these derive their essential importance, not from their own achievements or greatness, but from the extent to which these are consistent with the welfare of the whole body.

During the period covered by the table, the percentage of members of co-operative societies to the whole population of the United Kingdom increased from less than 2 to about 5, and the total membership in Great

GENERAL COMPARATIVE TABLE *

Year	Population of Great Britain	Co-operative Associations for Retail Distribution in Great Britain †					Income-tax assessments in Great Britain (£1000's)	Exports and imports, United Kingdom (£1000's)
		Members	No. of societies making returns	Capital (£1000's) Share	Capital (£1000's) Loan	Sales (£1000's)		
1881	29,807,211	546,712	964	5,377	671	15,392	549,113	694,105
1886	31,819,979	773,631	1139	7,913	956	20,375	593,096	618,530
1891	33,122,064	1,043,634	1298	11,309	1195	30,568	660,653	744,554
1896	35,057,011	1,354,021	1414	15,379	1515	36,622	646,110	738,188
1901	37,105,143	1,789,658	1423	21,952	3320	52,651	799,316	869,854
1902	37,528,925	1,889,178	1440	23,150	3534	55,209	832,643	877,630
1903	37,957,561	1,982,510	1437	24,194	3750	57,373	845,062	902,973
1904	38,391,090	2,072,792	1435	25,115	3956	59,162	867,666	922,054
1905	38,829,580	2,146,242	1430	26,047	4163	60,902	876,692	972,616

* Compiled from Statistical Abstract and Board of Trade returns

† Only the figures for Great Britain are given, as it is here that the strongholds of these associations are found. In Ireland, the centre of agricultural co-operation, there were, in 1905, only twenty-two retail societies, with 6773 members, share and loan capital £36,213, and sales £185,438. Even these small figures represent a nearly tenfold increase since 1881.

Britain in 1905, on the basis accepted on page 70, would have represented about 7·76 million persons, or about 20 per cent. of the total population. The total sales in 1905, adding on £185,438 for Ireland, and accepting the official estimate that the total wages bill of the whole country amounts to about 725 millions would have been equivalent to a little more than 8½ per cent. of this great sum.

CHAPTER III

CENTRALIZATION

The Co-operative Union—A Danger—Co-operative Wholesale Societies—Federation—The English Wholesale Society—Facts and Figures—Functions—Special Competitive Advantages of the Wholesales—Position of Employees

SO far, co-operative societies have been described for the most part as though they were detached units of distributive enterprise and social life, but at the present time very few are such.

Through the Co-operative Union the great majority of them are, as we have seen, connected with the chief propagandist body of the movement; and to the network of its organization, which the Union spreads over the whole country, reference has been already made. For purposes of guidance, legal advice, conference, the circulation of co-operative publications, and educational propaganda—to name some only among the objects which the Union sets before itself—the societies are provided with a rallying point and a nucleus that serve to give them a consciousness of unity that, amid the differences of aim that may be represented by the constituent societies of the Union or by the delegates sent up to its annual Congress, brings at times an invigorating sense that the movement is representative of hopes far larger than can be focussed in any single enterprise. And stimulus of this kind is of the greatest value, for the intellectual and moral difficulty of the movement is to keep the imagination alive and the outlook comprehensive. As years go by, this difficulty has been increased by prosperity and by the absorption of so much of the energy of the movement in the management of materially successful undertakings. It remains, therefore, in the altered and far more favourable conditions of to-day, for the Union perhaps more than for any other co-operative organization, to fan enthusiasm, to discover and to inspire gifts and powers of leadership, and to cherish the memory of great traditions, for the danger of co-operation appears to be to-day even more than it was twenty years ago, lest, as Arnold

Toynbee said at Oxford in 1882, "material comfort may diminish spiritual energy." *

The danger is, it is true, by no means peculiar to the co-operative world. It is, perhaps, national, and, at any rate, save among the very poor, it threatens large sections of almost every class. It is to the honour of the co-operative movement that in its case, in the midst of so many entirely admirable manifestations of its vitality, this danger should loom so large.

The explanation is found in the high aims with which the movement is traditionally identified, for in every relationship we grade our judgments. The dignitary of the Church, for instance, who is guilty of grasping conduct, is thought of more harshly than is the money-lender detected in the same offence, and so, in the co-operative movement, we look for something more than just dealing, capable management, financial success, and the exercise of thrift. It is thus that we are led to look to the Union for an impulse that shall attach as much importance to fearlessness and to progress as to safety, and that shall direct the

* "The Industrial Revolution, etc.," p. 230.

co-operative mind unceasingly from achievement to aspiration.

The close association that is now maintained with the Co-operative Wholesale Society makes it more necessary to emphasize this need than it would have been in the past, because the Wholesale, in virtue of its functions, stands emphatically for the industrial and financial stability as distinct from the moral energy of the movement.

The present Co-operative Wholesale Societies of England and Scotland are the sequel to earlier attempts made to provide the retail societies with the facilities for advantageous buying that a wholesale source of supply, controlled by and managed in the interests of the societies, can furnish. In 1832 the Owenites had made an attempt at Liverpool; in 1850 the Christian Socialists one in London, and in 1855 the Rochdale Pioneers had started a wholesale department. All of these, from one cause or another, failed, but their failure helped to prepare the way for the remarkable business success that has attended the efforts of the existing societies.

They, and especially the English society,

are now emphatically the big things of the movement in this country, and, at the present time, out of the total purchases of the members of the stores, goods to the value of about 27 millions are supplied to them by these great societies.

It is, however, rather in their constitution than in the magnitude of their operations that the essential cause of their importance is found, since, membership being confined to co-operative societies, and the societies not only elect ing the general committees of the Wholesales upon which executive power is devolved, but also sending up their representatives to quarterly meetings, to which the executives report, the whole of the federated stores have a direct corporate responsibility for every important step that is taken and for the policy adopted. In the case of the English Wholesale Society, the quarterly meetings of delegates are held at the Newcastle and London branches, as well as at other centres, in addition to the final meeting at the headquarters in Manchester. The final meeting is fixed for one week after the others. The delegates appointed by the societies can attend any one of the various meetings they select. The same agenda is considered at all the meetings, and the final decision in any controverted matter is arrived at by totalling the votes of the delegates wherever recorded. A quarterly meeting at head-quarters is generally attended by about 700 delegates.

This diffusion of the sentiment, and in many respects of the reality of responsibility, is one of the advantages that a federated industrial democracy can secure, and up to the present time the growth of the Wholesale Societies is at once a proof of the power they possess of adapting their methods of management to fresh needs, and a testimony to the soundness of their constitution. Started in 1864, with a constituent membership of the share-holding societies of 18,337, and sales at the rate of about £90,000 per annum, in 1905 the English Wholesale had a membership of societies that embraced over 1,600,000 members and sales that amounted to nearly 21 millions. In the same way the sales of the Scottish Society had increased from £81,000 in 1869 to nearly 7 millions in 1905.

It is no wonder that a great heave of

satisfaction, if not of pride, breaks from the co-operator as he thinks on these achievements, and the feeling is apt to be fostered still more as he walks through one of the great entrepôts of his society, or even as he turns over the pages of the Wholesale *Annual.* In this substantial volume he is glad to see particulars of the head-quarters' establishment of the English Society at Manchester, or of the great branches at Newcastle and London, and to learn that at these three distributing centres alone there are 3860 persons employed; to know that there are depôts at Bristol, Cardiff, and Northampton; special purchasing depôts, among other places, at Liverpool, Tralee, New York, Montreal, Copenhagen, Hamburg, and Sydney; salerooms at some of the more important centres of population in this country, where the Wholesale is otherwise unrepresented; shipping offices at Goole, Garston, Rouen, and Calais, and a small co-operative fleet of eight steamships. At these various centres he learns that 5794 persons in all are employed, and sees that there are 9352 at the various productive works, bringing the total employees of the society up to the imposing total of 15,145. If he pursues his quest, he will find that the *Annual* gives, on their smaller scale, similar particulars of the sister society in Scotland, the roll of the employees of which has now reached 6635.

Of the various properties of both societies illustrations are given, and the member of the humble village store can see pictures of the great ranges of buildings in which, as a member of a federated society, he has some feeling of proprietorship—the great offices, warehouses, depôts, and factories, and his imagination is perhaps especially stirred by the pictures of the distant tea plantations and the factory that the two societies jointly own in the Far East, and by the photograph of the great group of coolies employed thereon.

If he turns to the financial statement, he has little difficulty in seeing the results of the management of the committees in whose election he had a voice. Nearly every department, distributive or productive, is shown with its separate working, and he can generally see almost at a glance where losses, if any, are being made, and where gains; where ground is being lost or won. He can see swelling totals of aggregate profits (which it may be noted, although ordinary wholesale prices are charged, are, in proportion to turnover, not large) of membership and of capital, and, if he will, he can turn from the business record of current and past years to a series of general articles on wider matters of social import.

It is perhaps especially by the insertion of these articles that the general reader is reminded that he is studying the particulars of an organization that is identified with something more than ordinary trading and manufacturing enterprise, but it is only by recalling to his mind the million and odd members of the constituent societies, and remembering that these are members of the working-class community, that the really significant fact is brought home to him that the Wholesales are first and foremost demonstrations of the power of "labour" to conduct on a representative basis a great business and financial undertaking.

In various subsidiary ways, besides those already mentioned, the societies now make themselves felt. The English society, for instance, sends duly qualified auditors to audit the accounts of some eighty of the retail societies—a function that is probably destined greatly to increase in importance. "The Wholesale" is also a centre to which the committee of a store can turn for practical advice in starting business or in opening a new department; in buying land or in building new premises. "If capital is urgently wanted for some development, and a loan or mortgage is to be effected,"* a society is very likely to turn in the same direction. About 400 societies use *The Wheatsheaf*, the illustrated paper issued by the Wholesale, adding to it their few pages of local matter, in a way that is analogous to that adopted by so many parish magazines. By this plan a monthly circulation of 300,000 copies has been secured for a paper that, containing material of general interest, has it as a chief object to keep the claims of the Wholesale and of wholesale co-operation before large numbers of co-operators throughout the country.

It would, however, be erroneous to draw any very far-reaching conclusions as regards the future from the admittedly great positions

* Cf. "The 'Wholesale' of To-day" (1904), p. 38.

that the Wholesale Societies have won, or from the analogous examples of the great distributive retail stores, for in both directions, in as far as business capacity has so far been demonstrated, it has shown itself, considerable though the co-operative movement is, in rather limited and not very elastic fields.

As regards the Wholesale Societies, it has to be remembered that their economic position is very greatly strengthened by the fact that they follow the formation of well-disposed groups of probable customers. If pioneer work has to be undertaken, and a "connection" among consumers formed, this has probably been the preliminary task of some inner circle of the members of the various federated stores. When, therefore, such a store joins one of the Wholesale Societies, it does so as a centre of discovered demand. The Wholesales, although, in conjunction with the Co-operative Union, they undertake a certain amount of propagandist work, have thus never to "make trade" in the ordinary sense of the words, and the nature and the risks of their task are thereby immensely simplified and lessened.

Moreover, in dealing with their customers—the societies in membership—their position is greatly strengthened over that of competing wholesale merchants by exactly the same sentiment that *cæteris paribus* (which for many co-operators may be interpreted "prices elsewhere not dissuading") creates an exceptional inducement to individual members of the distributive store to deal with it in preference to going elsewhere. Thus, the investment of some of the store capital in the Wholesale, and the knowledge that it is being used in the interests of all the federated societies, and further, the fact that in the balance-sheet of the store the dividend paid by the Wholesale will figure at the end of the half-year, are each of them considerations that will weigh with the committee of the federated society, and still further strengthen and differentiate the position of the Wholesales as compared with that of the ordinary merchant.

When to such advantages there is added the convenience to the store manager and committee of being able to buy nearly everything that is required, groceries and drapery, no less than boots, chairs, and tobacco, from a single centre, that gives, as a matter of course, and without bargaining, good average commodities at average prices, the strength of the position of the Wholesales is still further realized.

The claims, in short, that these societies make upon the capacity, energy, and initiative of those who hold the most responsible positions in them, differ in many essential ways from those that press upon the heads of competing capitalistic undertakings, for the great federated membership of the Wholesales constitutes in itself a market ready to hand, such as no other sources of wholesale supply, unless for goods manufactured under monopolistic or semi-monopolistic conditions, can hope to possess.

The success of the Wholesale Societies is thus ultimately almost as much due to the same diffused sentiment that first led to the establishment of the local distributive stores as to great individual capacity.

That there is much room for this cannot be denied, and, if joined with the magnetic force of genuine leadership, far more scope, it may be, than is at present realized. With rare exceptions, however, the characteristic of those who have been called upon to take the leading parts in the organization of the Wholesale Societies seems to have been the possession of solid dependable capacity marked by the dogged persistency of those who find themselves on a good solid road and mean to keep it, rather than of those who are likely to be attracted to right or left by rugged winding paths, no matter to what shining heights these may seem to lead. The Wholesales are, in other words, the counterpart *en gros* of the vast majority of the rank-and-file of the movement at large, and those who control them are thus the selection that would be naturally made under a system of trading the administration of which is based on the political principle of representation.

Although the inherent business advantages that the Wholesale Societies enjoy, tend as we have seen, to diminish the difficulty of management and the scope for initiative, the success of these great institutions remains nevertheless a remarkable triumph of democratic business organization.

Although forming at the present time the

backbone of the federated store movement, the Wholesale Societies were originally started with much humbler purposes in view: to make the retail societies independent of wholesale dealers, who, under the jealous interference of retail rivals, were sometimes constrained or tempted to refuse to supply the distributive stores; to make it easy for the managers and committees of stores, often ignorant of business practices and perplexed by the varying conditions as to price and quality in the wholesale markets, to buy the goods their members required; and, lastly, to secure the profit of the wholesale trade. Two other objects that have gathered weight as the Wholesales have prospered have been to provide for the use and custody of surplus store capital, and to create a great machinery, not only of co-operative supply, but also in so far as possible, of co-operative production, by which the whole co-operative movement could be welded into a single federal system.

In the way in which this last aim has been pursued we reach one of the great sources of division in the co-operative world—of heart-burning in the past, and, perhaps, of serious practical problems in the future, arising from the obvious dangers of over-centralization. Other questions of principle also supervene, for the English Wholesale Society, as a considerable employer of labour that tends constantly to increase in volume, has made no effort in recent years to differentiate the position of its employees from that of those working under ordinary employers, who pay fair wages and whose treatment is considerate. Thus the status of the co-operative employee is left by this society exactly the same as that of most other wage earners, save that, if a member of a federated society, he secures as a consumer, his somewhat infinitesimal share of the dividend that the Wholesale may earn.

By the Scottish Wholesale Society a different policy has been adopted, and its employees, in virtue of their being such, have "shared profits." To the vexed questions thus raised we must return when something more has been written on the different way in which co-operative associations of various kinds play the part of employer, and on the analogous ways in which the relations of wage earners and wage payers are or may be arranged in non-co-operative undertakings.

CHAPTER IV

THE CLAIMS OF EDUCATION

Ignorant Spending—Educational Grants—The Social Fringe of Education—Music—Conflicting Aims—Methods adopted—Scholarships—A Special Claim

MUCH that has been written may seem to point to no very lofty achievement on the part of the co-operative movement, and, when judged by the highest standard, this is true, for it must be admitted that the largest and most widely spread results of the movement have been the diffusion of conditions that make simply for material comfort in life. Families, however, like armies, walk upon their stomachs. To fill these, and to secure their being filled even in the leaner years, is still the first step that many families have to take, even among co-operative households, and the co-operative movement has done much to see that these rudimentary ends are attained. Good and wholesome food and decent homes

—these are for all the first essentials, and because all men need them we are, perhaps, too apt to think that their acquisition requires little thought, and that they are certain to be ours if only their money cost be available. It is far otherwise, and it may be that one of the most important of the educational tasks that co-operators could set before themselves would be to carry out a propaganda on behalf of wise expenditure.

From early days "education," in the more conventional meaning, has been very generally accepted as one of the ends that co-operators should place before themselves, and year by year grants, often the 2½ per cent. of profits suggested by the model rules of the Co-operative Union, are made for this purpose by many of the societies. In the year 1905-6, about £83,000 was thus allotted.

A considerable portion of this sum goes, it is true, in teas and entertainments, and, it appears, in the circulation of co-operative literature that is very largely used for trading purposes. Much goes, therefore, in various ways that can be said to touch even the social fringe of education but lightly.

Even such uses of educational grants are, however, not to be despised. A "Co-operative tea," for instance, will often prepare the way for other forms of association that will carry members beyond the pleasures of mere social enjoyment. At the least, a common meal makes for common intercourse, and intercourse is the first step towards practical fellowship.

Co-operation is one of the few "movements" that, outside the religious organizations, gives prominence to song and to music, and the hearts and minds of co-operators of all ages might well be still further turned to the power of these, as a solace, an inspiration, and a joy. It is noteworthy that one of the great co-operative events, arranged each August at the Crystal Palace, in especial connection with the federated productive societies, is called, not a "demonstration" or an "exhibition," or even a "congress" although partaking something of the nature of all of these, but a "festival"—a glad and good name that it is well for England to have thus prominently associated with a phase of its industrial life.

Each year tends to change the nature of the appeal that education in the ordinary sense of the word makes upon co-operators, and the various forms of the public provision that are now made have led some of the societies to doubt whether co-operative funds need be any longer voted for these purposes. The story of the struggle of the Rochdale and other early societies to establish their library and evening classes, or to maintain "a school for young persons at a charge of *2d.* per month" is, for instance, not likely to be repeated in the experience of any future society. In these days of public libraries and reading-rooms, of evening continuation classes, of polytechnics and technical institutes, and of democratic universities, we are reminded of the advance, so far as organization and supply are concerned, that has been made, and find the explanation of the widespread opinion among co-operators that their best educational work will in the future be done by fostering the public provision, and by serving on public administrative boards, rather than by giving direct financial assistance to educational effort.

Although there is much justification for this opinion, since it is useless for a private corporation to undertake duties that are now generally and, at any rate, in centres of large and vigorous population, placed in the more efficient hands of the public authority, certain fields remain open that can still be best filled by organized co-operative effort, and that, from the co-operative point of view, are never likely to be adequately filled in any other way. I refer especially to teaching in the history and aims of the co-operative movement itself, and in those chapters of economic history in which it has played and seems destined to play a not inconsiderable part. Such instruction would probably include the technical parts of co-operative work—the details of organization, rules, privileges and conditions of membership, etc.—but, if rightly conceived, would go far beyond this, and especially when a responsive class had been formed under the direction of a far-seeing teacher, it would be by such classes that the highest hopes and noblest aims of the movement could best be kept alive.

There is, perhaps, a certain danger that has to be avoided, when the direction of such teaching is vested in those who are within the ranks of the movement itself—the danger of not attaching sufficient importance to the right reading of the industrial life of the community at large. Enthusiasm inspires, but it is apt to darken counsel, and one of the most difficult tasks of co-operative teachers is to appraise at their true worth those elements and forces of industry that lie outside of a movement to which, it may be assumed, they themselves attach supreme value. In reality, the tasks of co-operators—even in as far as they are concerned with the betterment of the conditions of industrial life, are being undertaken in many directions—and as an economic force co-operation is in itself quite unable to meet the demands that the complexity and urgency of modern trade and industry enforce. Nor is it able to adapt itself to the temperaments and impulses of many good men.

Co-operation must, therefore, be recognized as part of a life that is far greater than itself—the great scheme that was never planned, but which grows by a manifold energy—of bankers, manufacturers, merchants, dealers, carriers of every kind, operatives, and labourers—the business life of the nation. In this life co-operation takes its place as one only of the forces that must be used and relied upon to secure the spread of well-being and well-doing, and it is this wider view that places co-operation in its true proportions, that the best and most hopeful co-operative teaching will make possible.

Although, therefore, co-operation is but a part of the Gospel of Industry, those among its missionaries will be the most powerful who cling, in spite of every temptation that the lower success of the movement may seem to offer, to the noblest traditions of the spiritual fathers of the movement. This inheritance of teaching and example is a great responsibility, and lays those who ignore it open to the ancient offence of sinning against the light. But, on the other hand, it is a great reserve force helping to safeguard the movement against degeneration when the downward forces threaten, in the one direction of exclusiveness or self-gratulation, in the other of impatience or envy.

Thus, whatever provision may be made by others, the obligation to teach the technicalities, the history, the relative position, and the aims of the movement will remain, and in this something has already been attempted, nearly 9000 students, for instance, having been enrolled in the year 1905–6 in the classes arranged by the various education committees. The ordinary lending library, making provision for the normally preponderating demand for fiction, is likely to be less frequently found in the future, but small, well-selected libraries of economic and sociological literature are likely to increase rather than to diminish in number and value, while the co-operative reading-rooms, of which about 400 are now in existence, cannot be diverted from their club value even if public provision makes them less important in other ways.

In directing the educational efforts of the co-operative movement, the Co-operative Union exercises considerable influence, and the establishment of two scholarships at Oxford is an interesting sign of the times, even though the way they have been used does not always appear to give complete satisfaction. The scholarships are memorials to Edward Vansittart Neale, late General Secretary of the Union, and the man to whom, Robert Owen apart, co-operators owe more, perhaps, than to any other single individual; and to Thomas

Hughes, and are now open to sons of members of co-operative societies affiliated to the Union.

Prizes are not infrequently offered by the Union and the Societies for essay or other educational competitions, but it is probable that the practice of granting small scholarships similar to those, for instance, of the late Technical Education Board for London, and on the lines that have been already offered by some societies, might be often adopted. The cost need not be very great, but the impulse might be, since the qualification might be so arranged that the benefit of the competition would be shared by large numbers. The proposal to pay school fees or part fees, if carefully organized, would have similar social value, for it is just among such a body as that of the co-operators that there must be many young lives, boys or girls, by the prolongation of whose years of training both they themselves and the community at large would greatly gain. Among the scholars thus aided, who would probably use their grants, whatever form these took, in some local secondary or technical school, a natural development would be the selection of a few of the most promising, and the provision by one or another of the central organizations of more University Scholarships, some of which might be tenable, not necessarily, although perhaps preferably, at either Oxford or Cambridge. In other cases the scholarships might with advantage be attached more locally, and the ties between the sectional organization of the movement and the more modern territorial universities of Liverpool, Birmingham, Wales, or elsewhere, be strengthened.

In such and analogous ways, as through occasional lectures and conferences, the educational activity of the movement is fairly assured, because of its traditional and genuine interest in education, and because as an organized body controlling considerable sums, developments of this kind are easily practicable.

In other directions, in simpler and more domestic ways, the educational tasks of co-operators can be pursued with more exceptional chances of widespread effectiveness, because of the facilities that the general organization of the movement gives of spreading information; because of the close connection that normally exists between the stores, through what they sell, and the home conditions of the members, through what they buy; because of the large number of women who are attached to the movement, and of the increasing number of these who are becoming interested in its welfare; because of the effective organization that women are developing in their Guilds, and because, side by side with all this organization and growing interest, there exists so much ignorance in matters affecting vitally the happiness of the family life and the health of its members. It is thus among women, and through women, that some of the most distinctive and most valuable educational work of the co-operative movement is being and can be done. In effort so directed there is no smallness of aim, for as its ends are attained, so also are the surest foundations being laid of the whole structure of healthy communal life.

CHAPTER V

A QUESTION OF "CLASS"

Mainly Composed of Weekly Wage Earners — Salaried Classes Hold Aloof—Probable Future Developments

THE facts and figures that have been given show that so far as the co-operative movement has developed up to the present time it has depended very largely for its adherents upon those who are in regular employment, using this expression in the relative and qualified sense that the conditions of modern industry impose,—upon those among whom there is a great solidarity of interest, and who by the conditions of their employment are brought together in large numbers, and among whom therefore not only are the chances of isolated industrial effort small, but among whom the opportunities for the formation of common opinion, of common interest, and thus of common action are great.

Since the stores are democratically controlled, a certain amount of public spirit has been demanded alike from those to whom responsible positions have been given, and at any rate by an inner circle of those by whom this responsibility has been delegated.

When there has been no co-operative tradition, something of the co-operative temperament has been necessary—a quasi-political attitude of mind directing itself to questions of industrial and social well-being. Finally, since the financial basis of the society has been the possession of at least a small margin of reserve capital by the individual member, even if this was only represented by the shilling paid on application for membership, and by the willingness to let subsequent dividends accumulate to the price of the qualifying paid-up share, the constitution of the society has been unable to appeal to those to whom the idea of even the smallest margin of reserve is quite alien, and who, moreover, in spending habitually think only of the return of the moment, and whose minds therefore never, and whose purses perhaps not very often, are able to reach out to the idea, or the acquisition, of a deferred advantage.

Thus, from various causes, social, economic, and moral, the co-operative store movement has run its sturdy course almost exclusively in the fairly well-defined channels of the "better sort" of working men and women—artisans and mechanics, miners and mill-workers, and those just above or just below the classes thus roughly indicated.*

The lower-salaried classes have on the whole held aloof even when their own occupation has had nothing to do with competing branches of distributive industry, and when, therefore, some bias against the movement might have been expected.

The difficulty with which the working man associates with those who are very often not so

* The class of member bears in an important sense upon the question of the liability of co-operative societies to income tax, a matter that again assumed prominence in 1905 in view of the reference by the Chancellor of the Exchequer to the Income Tax Committee of an instruction "to inquire and report whether co-operative societies enjoy under the present law any undue exemption." The following extract from their Report will show the view adopted by the committee on this point. "We also think it clearly established that, in a society of the Rochdale type, the dealings of the society with its own members do not result in anything which can be treated as 'profit' within the meaning of the present Income Tax Acts, or which in fairness could be so treated under any amendment of the law" (p. 24, § 135). According to the secretary of the Co-operative Union, "it is well known that the bulk of the membership of co-operative societies—at least, 95 per cent.—consists of persons who are not in receipt of taxable incomes" (see "Co-operative Societies and the Income Tax." By J. C. Gray. 1903).

well off as himself, either nominally or really, but who are not manual workers, and who, in a sense that is often symbolized by the obligation to wear a black coat and a white collar every day, have "to keep up appearances" in a sense from which the weekly wage-earner and manual worker is comparatively free, is illustrated afresh in the experience and history of the co-operative movement.

"Birds of a feather flock together," although generally quoted with a sinister meaning, is a proverb of great social significance, and its practical truth may be observed in a hundred ways in the experiences of everyday life, from the arrangements made or in practice enforced by most of those who are responsible for the institutional religious life of the community to the little group of men that meets together as in a club in the bar-parlour of some quiet out-of-the-way beerhouse.

The magnet that draws like to like may be charged with an attractive force, that it is the outcome of the highest sensibilities, moral, intellectual, or æsthetic; or it may be the expression of a crude selfishness that, different in form, is common in its essential features to large numbers of almost every class; or again, and perhaps more commonly, it may be the result of genuine differences, not so much in standard of life as in felt needs, in modes of enjoyment, in speech, or in manners.

It is not exactly rational, because it is very largely traditional and inherent; but, on the other hand, it is not a sign of weakness, or pride, or of selfishness. It is the practical recognition of real and important differences.

Something of this kind is probably at the root of the slow progress that co-operation is making among those who do not belong to the "working classes," and when we are told that the barrier is due to a sense of "superiority," this explanation is expressive only of a half truth.

If the classes "above" felt that Industrial Co-operation could offer them advantages that would be as real to them as to the rank and file of the existing members, there is no reason to think that these advantages would not be secured, and the "upward" borders of the movement be rapidly extended.

If, however, any considerable extension in this direction comes, as among the classes roughly indicated by second-division civil servants, clerks, and elementary school teachers, it will probably come through the unprompted desire of these classes themselves, rather than through any active propaganda, for the motive to this is to a great extent lacking. Genuine propaganda, in connection with such a movement as that of the co-operators, in spite of all the charges of selfishness that are levelled against it, is directed primarily where it is felt that those outside the movement will benefit most from the adoption of its principles, and where, therefore, what is regarded as the need for co-operation, is most marked. Thus, next to the great field that is still untouched by the movement, on the present plane of its development, it is felt that "downwards" rather than "upwards" (to use these offensive but convenient words), there is the more urgent need for expansion, for, as we have seen, up to the present time, co-operation has left "the poor" almost entirely untouched. In this connection, the most significant attempt is being made in close connection with the women of the co-operative movement, and to a consideration of their organization and their work we may now turn.

CHAPTER VI

CO-OPERATION AND WOMEN

The Women's Guild—Its Constitution, Aims, and Uses

NOT only is the Co-operative movement composed of men and women, but the latter, in the case of some societies, form the majority of the members, and, in the movement as a whole, an important minority. Large though the number of women members is, their importance is not to be measured in this way, for through their "basket power" it is on them, whether qualifying personally as members or not, that the stores are practically dependent for their existence.

Even simply because of their power in the movement, it was natural that women should wish to possess some organization of their own, that would give them not only fresh interests in life and, as need might arise, the chance of making themselves heard, but also opportunities for associated action on behalf of the movement itself, and especially on behalf of women, girls, and children. The beginning of such an organization is found in a modest suggestion made by Mrs. Acland at the beginning of 1883, in the "Women's Corner" of the *Co-operative News*, that "co-operative mothers' meetings" should be held, "when we may bring our work and sit together, one of us reading some co-operative work aloud, which may afterwards be discussed." *

This was the inception of the movement, which speedily led to proposals, formulated in the first instance by Mrs. Lawrenson, of Woolwich, for the formation of a central organization with local branches, local officers, and the dignity of a distinctive name.

The hopes then born have never been allowed to die, and the Women's Co-operative Guild, from the 235 members and six branches which had been formed by the middle of 1884, is now a large federal body, composed of 452 branches, divided into "sections" and "districts," holding its own annual congress, and comprising a membership of 24,000 women.

"The principle underlying the whole

* Quoted in "The Women's Co-operative Guild," 1883–1904, by Miss Llewelyn Davies, p. 10.

organization of the Guild from top to bottom is that of self-government," and it forms a genuine "democracy of working women." * The analogy of the "mothers' meeting," adequate perhaps at the outset, speedily give way as wider projects made themselves felt, and at the present time, the Guild concerns itself with the largest questions, co-operative and even national. Its place is fully recognized by the central organizations and the chief executive bodies of the movement, and in 1904, 374 representative positions of one kind or another outside the Guild itself, some of them of high responsibility, were filled by women. "Co-operative work has two sides to it; there is the business work and there is the social and educational work," says Messrs. Acland and Jones, in their little book on "Working Men Co-operators," and it is to the second division that the most fruitful efforts of the Guild are directed. It is noteworthy, for instance, that while only 35 women are members of management committees (in 23 different societies), 286 sit on educational committees (in 116 different societies).

"The Women's Co-operative Guild," p. 35.

Marked though the success of the Guild has been, it has not been uniformly won, and the break-up of branches that had been formed has often happened. Sometimes this has arisen from lack of sympathy and financial support from the societies with which they have been connected, and sometimes the responsibility for failure has rested with the women themselves. "It has been very difficult in some cases to interest the women in Guild work."* Existing branches, moreover, vary greatly in size and vitality, and the following is given by Miss Llewelyn Davies as the picture only of what would be seen by a visitor to those that are well supported and active :—

"We should see," she writes, "a company of women in a fine co-operative hall over the stores, presided over by one of their own members on the platform, the secretary seated by her side, with minute-book on the table. Business will be first disposed of, for example, correspondence with the general secretary, suggesting an anti-credit campaign or a Factory Bill resolution; the election of delegates to a conference or a congress; arrangements for an excursion over a co-operative

* *Op. cit.*, p. 38.

factory, or with the education committee for a Christmas or summer treat for members' children. Then may follow a lantern-lecture on the Guild, Co-operative Productions, or 'The White Slaves of England,' or a visit from one of the Guild officials, or a County Council lecture on sick-nursing, or an address from a committee-man of the store on the balance-sheet, or from a Poor Law Guardian, or the Secretary of a Woman's Suffrage Society. Or the subject of the evening may consist of a paper on Co-operation or the Guild by a local member, followed by a discussion." . . .*

Not infrequently the women now speak at or preside over meetings of a more public character than that just described, and on such occasions husbands often learn the unsuspected powers that their wives possess. Thus, at a recent conference, over which a Guild-woman was presiding with quiet dignity, her husband remarked, "I have known my wife for sixteen years, but I never knew until to-day that she could do *this*," with an emphasis on the last word that was no less expressive of pride and pleasure than it was of surprise.

* *Op. cit.*, pp. 39, 40.

The extract given above indicates the wide range of interest shown by the members of a vigorous branch, but it does not reflect the personal satisfaction, that is not the least of the benefits which the Guild ensures, for the correlative effect of taking interest in such subjects as those mentioned is a greater zest in life. "Home is best, but its duties sometimes pall, especially if unrelieved by any change, and for many women, no evening is said to be more welcome than that of the Guild meeting—interesting, often stimulating, and always friendly. This is how Northumberland Guildswomen speak of the meetings—

"The Guild night is my rest. Ah hed been weshing arl day, an' was hurrying to get ready. Wor man says, 'Ye better sit yoursell down an get a rest.' Ah says, 'No, ah rest at th' Guild, an hear somethin' gud inta the bargain.'" . . . And another, "Ah have been bad arl day, but ah knew if ah could get to th' Guild ah shood be a bit lot better."

"Few men," says the historian, with truth, "can realize how much drudgery and lonely effort there is in the everyday work of a housewife. If men need a life apart from their daily work, women need it equally." *

To those women who take more active and more responsible parts in the work of the Guild come other influences, making sympathies more real, forcing them to think out questions for themselves, and, in order to save time, that has suddenly become more precious, obliging them to be methodical, and to avoid its waste.

"One of the things the Guild teaches is system," remarks a member. "To be able to attend branch meetings and conferences, and do your household duties, you must have system in your home work. You can't loiter over it." †

And another member, who is also a Poor Law Guardian, writes:—

"I have had a splendid constitution, and the busy life suited me. Most of my lectures and addresses have been thought out when my hands have been busy with household duties—in the wash-tub, when baking bread (and, by the way, I have never bought a week's baking during my married life of over twenty-one

* *Op. cit.*, p. 151. † *Op. cit.*, p. 114.

years) or doing out my rooms. Somehow the work passes more quickly, and I have not felt the work so hard when my mind has been filled with other things." *

* The following are some specimens of the "system" adopted by Guild members :—

Monday—Washing and ironing.

Tuesday—Tidying, sewing, preparing dinner for Wednesday.

Wednesday—Poor Law day.

Thursday—Bake and clean bedrooms.

Friday—Clean rest of house, and have help for half a day.

Saturday—Two or three hours of Poor Law work.

The following is the order of her days adopted by a Southerner, living on the outskirts of a town in a six-roomed house—

Monday—I get up at twenty minutes to six, and light fire in kitchen; then light fire for copper and fill it with water. Then cook bacon, put dinners ready, and start off three [members of the family to their work] by twenty-past six. Then, of course, water is warm ready for washing. Do not clean up house except breakfast-things, till clothes are drying. Fold clothes, and come out to nursing-class. Sometimes I can get mangling done in afternoon, if the clothes dry quickly.

Tuesday—I iron up till dinner time, make pies and cakes for men's dinners—jam-pies, bacon-pies, rabbit-pies, meat-pies. After dinner I make cake for breakfast, cake for their dinner, cake for their tea; they'll eat cake nearly always.

Wednesday—Finish my ironing, if I had extra lot, till dinner. Sewing meeting, or running about for Guild.

Thursday and *Friday*—Turn out bedrooms, or a good big day of sewing. Stores committee meeting on Thursday, Guild on Friday.

Saturday—Touch up all round; hot dinner at one thirty. *

* *Op. cit.*, pp. 151–3.

CHAPTER VII

HEALTH PROPAGANDA

Importance of Good Cooking—Cleanliness—Sick-nursing—Value of Women's Guild in Spreading Knowledge of Health Conditions

WIDER interests, much personal satisfaction and enjoyment, and, in many individual cases, improved home conditions, have resulted from the work of the Guild, and from the association of its members with the stores.

Many of those who join the Guild, and the stores too, are themselves wage-earners, but more often home is their workshop. It is the husbands and children who serve in the outside ranks of industry. Thus, in the vast majority of cases the women have some or all of the ordinary domestic cares upon them—of buying food, cleaning rooms, washing, baking bread (in the North), cooking, managing the children, and many other duties. The woman is domestic economist in chief. Upon her more than upon any other depends the character of the home.

And almost always, even in the best managed households, there is much to be learnt. In buying wisely and to good advantage; in the knowledge of the value of foods;* in good and methodical management; in averaging expenditure throughout the year; in knowing how to cook well, and economically and attractively; how best and most easily to keep a home clean, and how, with inexpensive devices, to make it pretty; in appreciating the importance to health of well-cooked food and regular meals; of fresh air by day and night, of the bath, and of suitable clothing; in the management of the sick-room, and in the care of invalids; in the arrest of threatening maladies, and in the management and feeding of young children—in all such ways, the community itself is still but half taught.

It is the apparently large subjects, upon which political or municipal or it may be voluntary corporate action can be taken, that are apt to be most attractive to women as

* Note the excellent suggestion that in the proposed "cooked meat shops" of "poor stores," "bills of cheap and varied diet for a week as suggestions for home consumption" should be posted up. (*Report on the Extension of Co-operation to the Poor made by the Committee of the Women's Co-operative Guild*, p. 34.)

to men. Such subjects smack of battle and struggle. They are more generally recognized as important. Power of some kind may come, or seem to come, from their successful pursuit. But, as compared with such matters as those just mentioned, they are often such as a modern Naaman would choose—the Abana and Pharpar of social progress, the real Jordan, or, at any rate, its major stream, often flowing unnoticed and unsought at each man's threshold.

In the conduct of what might grow into a beneficent health crusade, the Women's Co-operative Guild has already slightly moved, and lectures on sick-nursing and cookery are mentioned among the fixtures arranged by some of the earliest branches that were formed. The care of convalescents has become one of the minor objects of the Guild, and the co-operative movement, as a whole, is directing some attention to the same subject. In administrative work it is interesting to see that experience is said to show that "a point women often seize on more quickly than men is need for sanitary reforms." * In 1897 what

* "The Women's Co-operative Guild," pp. 101, 102.

almost amounted to a campaign on the Public Health Laws was undertaken, while the abiding interest shown in questions of factory legislation turns largely on the supreme importance of health.

There seems no reason, however, why its claims should not be more systematically and persistently advocated. If the need of health knowledge were more vividly grasped, this would probably be done, and co-operative women, helped by expert advice, might do much to promulgate, or help to promulgate, the new Mosaic code of health which the nation so badly needs. If such a body of women as that of the co-operators, independent and many of them now habituated to self-government, will demand this knowledge, it is far more likely to spread than if it be supplied on the initiative of those, who, already possessing it, are conscious of its importance. It must be asked for by the rank and file—not simply promulgated by experts, even though they be enthusiasts—if it is to win its way.

The history of the Guild is also often the history of success won against difficulty. The "hard crust" of enthusiasm, slow to move, and sometimes of downright lack of sympathy, has often had to be broken through. As a rule, however, objects aimed at seem to have been won, even though by importunity, as in the case of twelve Guild women who are reported to have gone to their store "in turn" "day after day," till they got some co-operative article which was not kept in stock, but which they were determined to have. If with like determination, these women resolve to have placed at their disposal, not the health knowledge of specialists, but such as can be easily applied and should be shared by all, which doctors, analysts, medical officers of health, sanitary visitors, district nurses, health missioners, and cookery instructors, possess, they would get it. The contents of the best and simplest leaflets and booklets of the National Health Society, of the National Society for the Prevention of Consumption, the leaflets dealing with the care of children used by some hospitals, and similar ones on cognate subjects issued by other agencies, the most assured results of chemical science on the dietetic value of ordinary foods, and the merits and demerits of the hundred and one patent products which are widely advertised and about which every housewife is apt to be perplexed, are suggestions of the kind of knowledge that might be easily placed at the disposal of every Co-operative Guild woman, and through the admirable organization of the whole movement, aided by the *Co-operative News*, its official organ with a weekly circulation of some 70,000, at the disposal of almost every co-operative family in the country, representing now, as we have seen, some 8,000,000 persons. If there be matters of controversy on great health issues, as indeed there are, the plan of sectional conferences, discussing if need be the same subject and the same paper in all parts of the country, would immensely facilitate the wide-spread consideration of selected subjects of importance.

In connection with health teaching, and the preservation of health as a personal, a civic, and a national obligation, the claims of the children are paramount, for the rates of infant mortality are to-day the great blot upon the vital statistics of the country. And the deaths that are recorded are but the index of the unnecessary struggles of thousands of those who live, permanently handicapped by the ignorance or carelessness of the guardians of their helpless years.

It would be an achievement of the greatest possible value alike for the co-operative movement itself and for the country at large, if on published returns, it could be shown that co-operative women, using the resources of their societies as centres of wholesome supply, forcing the matter upon the attention of every responsible organization in the movement, and giving exceptional thought and the most intelligent response possible to the dictates of that tenderness which is ready to knock at every mother's heart, were keeping the chill visitor from their homes with far more than average success. This would be a demonstration of the social influence and value of the co-operative movement, that would be far more telling even than great membership or great sales, and others would inevitably ask, What is this alchemy of care and knowledge by which these great results are won ?* Other

* Attention is already being directed to these matters, but incidentally rather than as a matter of prime concern. Signifi-

gains, moreover, will follow, when the youngest are nurtured, not simply with affection, but also with a fuller understanding of what the healthy body means to children in after life. The general health standard of the family will tend to be driven upwards, for it will be recognized that those of every age may and should share in this first blessing of full bodily development.

We touch here, in connection with young children, other questions of fundamental social importance, for in children—the prime products of the reproductive powers of the race—we see

cant in this connection is the suggestion made by a committee of the Women's Co-operative Guild that a possible development of the work of co-operation, in its more propagandist forms, might be the establishment of "a dispensary to which a certificated midwife may be attached." It is open to question whether it is desirable to multiply the facilities for securing medicine, but at a moment when the Act for the Registration of Midwives is coming into effective operation, there is probably in connection with it a considerable field for the exercise of the influence of the Women's Co-operative Guild. It could, for instance, perform admirable service in helping to raise the level of the calling of the midwife, and the tone of public opinion concerning it, so that not only may greater expert skill be looked for, and more uniformly provided, but that the poor may learn to expect to find in the midwife some of those higher personal qualifications by the presence of which alone can the best influences, hygienic and moral, be brought to bear at these times of crisis in a mother's life.

reflected more vividly than in any other form, the essentially dynamic nature of the problems that life and industry present. Large and ill-nurtured families already stare us in the face, and if the nurturing power be increased, so also, it may be argued, will the pressure of poverty itself; what the youngest gains, the family, it may be said, will lose.

It is an unproven and probably a fallacious argument, and is perhaps the ghost of a chill and discredited doctrine of the past, but it must suffice to point to the well-established fact, that the highest death-rate is often co-incident with the highest birth-rate, and to cite the reasoned opinion of Mr. Charles Booth, that—

"The death of a child, especially if it be a baby, does tend to bring about the birth of another. If child mortality could be checked the birth-rate would certainly be reduced, and a terrible waste of every kind would be prevented." . . . "On the whole," he concludes, "it may fairly be expected that concurrently with a rising standard of health we may see a fall in birth-rate as well as death-rate, and thus have no cause to fear as the result of better sanitation, that the largest natural increase in population will ever be contributed by the lowest class." *

But in truth, in every class the greatest by-product of the time that is waiting to be caught and transformed, is life itself, and health in the home is the surest fulcrum by means of which other gain, social and industrial, can be won. Care for the home is, moreover, the best starting-point for securing administrative measures of sanitary reform itself, for the lower the standard in the home, the lower is apt to be the standard outside, since public authorities tend constantly to drop towards and to excuse themselves by the level that local public opinion will sanction or tolerate.

Perhaps no organization has it so naturally within its powers to spread the knowledge among men and women by which health conditions can be secured or improved as that of the co-operators, and as regards all matters of domestic hygiene, few bodies could so effectively demand the dissemination of this knowledge, as the Women's Co-operative Guild.

* "Life and Labour of the People" (final volume), pp. 20, 26.

CHAPTER VIII

"POOR STORES"

Characteristics of Poor Districts—Social Disintegration—The Failure of Co-operation to reach "the Poor"—Recent Attempts—Recommendations—The Sunderland Experiment—The Store as a Centre of Social Effort

IN the light of such spheres of activity as those to which reference has just been made, and indeed in all its more effective and significant manifestations, Industrial Co-operation in Great Britain to-day is seen to derive its chief importance from the part it can play as a social influence rather than as an industrial system. The "store" supplies the master key, but the real sanctum lies beyond and above.

It has been for many years one of the regrets of large numbers within the movement, that "the poor" are for the most part without its borders, in the smaller class of town in which co-operation flourishes, hardly less than in the great centres of population where it experiences so much greater difficulty in securing a firm foothold.

There is much to explain this difficulty in reaching "the poor" besides the economic weakness and the frequent precariousness of their means of livelihood. In poor districts public opinion is apt to be limited to the smallest area—the court or the street. The bonds of any wider cohesion are looked for in vain. Religion, the conditions of employment, clubs for trade or friendly or social purposes—all alike fail to provide any effective bond. There is, in the limited area, much kindliness, often a great practical pooling of resources, and a greater community of goods than is shown by perhaps any other class, for none, impulsively, are kinder to the poor than are the poor themselves. But their kindness is accompanied by much disorganized living, by shiftlessness, and by waste. Any common movement is spasmodic, generally bred of momentary excitement or of exceptional distress.

Even among the more permanent occupiers therefore in such areas, neither the money resources nor the mental or moral basis are present in the quantity or quality that could make the task of any sustained organized movement other than one of exceptional difficulty. And many of the units change: some fresh faces constantly dropping down from the ranks above, others migrating for a season or for always. Many of the denizens of these districts prefer the casual mode of life. They have lost their birthright, but, as it did to Esau, disinheritance not infrequently brings some of the pleasures and excitement of the chase. An irregular mode of life has become a habit.

It is, however, just where this irregularity intensifies the normal difficulties of industrial life that it has been often felt that an exceptional chance may offer for co-operation to prove itself of special use, and from time to time efforts have been made to spread the movement in these difficult fields. In London, for example, an attempt was made a few years ago by the English Wholesale Society; local stores being then started and relieved of some of the responsibilities that the ordinary stores incur of management and of raising capital. In some towns branches have been opened in quite poor districts by well-established societies with a deliberately propagandist aim. These attempts have, however, never been very numerous, and they have never met with very great success. Co-operation has practically failed to reach "the poor."

During the last few years the problem has been receiving renewed attention, very largely through the efforts of the Women's Co-operative Guild, and an experiment made by the Sunderland Equitable Industrial Society—a body that has its own very honourable traditions of disinterested service *—in active association with the Guild, possesses various distinctive features of interest and value.

The attempt that the Guild has been making to throw light on this question of how best to spread co-operation among the poor has been seriously undertaken, and a systematic attempt was made to discover in what ways past experience seemed to show that there was the greatest chance, by adopting the sound test of business success, of putting the everyday life of poor families on a sounder basis. A careful investigation was

* Cf. "Fors Clavigera," lvii. Note on death of John Hopper.

made of the measure of success attained by co-operative societies that had already established branches in poor neighbourhoods, as in Bristol and Plymouth, and also of the special obstacles which the customary organization of the store seemed to place in the way of the poorer classes of consumers. The special attractions and features of non-co-operative sources of supply were also studied at Sheffield and elsewhere.

Various conclusions were arrived at pointing to the necessity of making things easier as regards the conditions of admission, simpler as regards withdrawals of accumulated savings, and more attractive as regards the goods supplied. A more novel and more distinctive conclusion arrived at was the desirability of providing living rooms for those whose special object was to be to carry on propagandist and social work; the store was to become the headquarters of a small "settlement."

The following are some of the more specific recommendations made by the special Inquiry Committee appointed by the Guild:—

That no entrance fees should be charged to new members of the stores.

That the payment of one shilling towards share capital should admit to membership.

That subsequent payments from dividend towards increase of share capital should not be enforced.

That when share capital, above one shilling, was owned, any sum, however small, might be withdrawn.

That dividends should be payable at any time, those claiming it before the declaration of dividends receiving sixpence in the pound less than the amount last declared.

That dividends should be paid on purchases of any total down to half a crown.

That penny banks for adults, as well as children, should be open, as far as possible, daily for deposits, and for withdrawals, every day at a central store, and on Mondays at the "poor stores."

That general clubs be formed, small subscriptions from a penny upwards being taken, and goods supplied at any time for any sum paid up.

That the most nourishing food at the cheapest prices and in the smallest quantities be supplied, a window of the grocery store to be "full of lowest priced goods and pennyworths."

That frozen meat and cooked meat shops be opened.

That coal-yards should be provided, where penny basketfuls could be obtained.

That milk should be sold in halfpennyworths or pennyworths.

Such were the recommendations, and it is clear that their adoption even in part would be calculated to dispel the widely-spread prejudice among the poor that the working-men's stores are not intended for them. It could no longer be thought that customers for small quantities were "not encouraged," or that the class of trade done was "too high."

As stated, it has been in Sunderland that, as the development of steps already taken, the most important move was made towards forming an adequately equipped "poor store," including the rooms for resident propagandist workers.

Hopes ran high among those who were enthusiastic for the new step.

"It was a red-letter day for Sunderland and for the co-operative movement," writes

Miss Llewelyn Davies, "when the corner premises of the branch store in Coronation Street were opened on October 8, 1902. The name 'Coronation Street' has become a household word in the movement, so great and widespread has the interest of co-operators become in the Sunderland experiment."

The following is the glowing description of the premises given by the same:—

"The bright, new store buildings consist of a grocer's shop, a butcher's shop, flour store, a hall, and small rooms for two resident workers. There are no shop windows in the street to compare with the co-operative ones. Plate glass and electric light show off the pile of co-operative loaves at 1*d.* and 2½*d.*; or an ingeniously arranged show of Co-operative Wholesale Society 1*d.* goods, pots of jam and marmalade, tins of tea and syrup, bottles of hair and castor oil, etc. Sometimes the cheapest drapery fills one of the windows, and a great bill announces a special sale in the hall, while various notices are pasted up announcing, it may be, the sale of milk at 4*d.* a quart, or the variety of meetings and entertainments going on in the hall upstairs.

"The butcher's corner window is equally attractive, with its steaming hot boiled joints of pork at 1*s*. or shanks at 5*d*., hot pease pudding with gravy ½*d*. and 1*d*., tripe 4*d*. and 5*d*., brawn 5*d*. The sale of hot soup at ½*d*. and 1*d*. was found so popular and such a financial success that it was introduced with equally good results into a neighbouring east end branch store." *

The district had been well if courageously chosen, for Coronation Street is a small shopping thoroughfare running through the midst of the most crowded and poorest parts of Sunderland, with "the highest record of death, disease, and crime." The men of the district are largely labourers connected with the shipyards; many of the women are hawkers of fish and other wares; the girls often find their employment in brewery or tobacco works; while many of the children swell the great army of sellers in the streets. It is enough to pass even once through the main thoroughfare and to glance down its lateral courts and ways, to be made conscious of the low local standard of industrial and social life.

In such a district, in which, in spite of its prevailing character, there is "much that is

* "History of the Guild," pp. 83, *et seq*.

good and attractive" to be found, the new experiment was started. The poor were met on their own ground, and their sympathies won, in the first instance, by the very practical step of offering them shopping facilities almost of their choosing. If the effort had stopped there but little good might have been effected. There are even elements of danger in simply multiplying the sources of cheap supply, and it is not always kindness that adds on a penny to the value of sixpence. Cheap food depôts alone, like easy-going almsgiving, may but increase the margin that is available for waste or for idleness.

In Coronation Street, however, it was more than a well-planned and carefully staffed store that was planted down. The rooms for resident ladies made possible that subtle personal element that money cannot buy, and that periodical visits paid by those who live at a distance can only in the rarest cases furnish. As you entered the grocery shop there was a desk behind which the "store ladies" were to be found every day at certain times "making friends with the people and doing business at the same time." Here new members were made, deposits were taken towards the shilling entrance fee and towards share capital, and pennies were put on club books, for the store was found to be "an admirable ground for propagandist work of all kinds."

Thus the hand that offered cheap and wholesome food of every kind, offered also friendship and neighbourliness. The store became in an exceptional way a centre of fellowship, and the hall was the scene of a succession of meetings, club gatherings, concerts and discussions, such as one might expect to find at an ordinary social settlement. Chances for saving were multiplied. The houses round about were visited on behalf of the store, or of the penny bank, or of the Temperance League, and for all such visitors doors were readily opened. Occasionally short excursions were arranged.

The experiment, on both the business and the social side, was successful, especially among women and children. During the first fifteen months, 303 new members joined, and although sales did not show progressive increase in every quarter, the volume was considerable, and averaged about £780 a month in the grocery and butchery departments. Those who foretold loss were thus agreeably disappointed, for the profits compared favourably with those made at the various other branches of the flourishing parent society, with its 17,000 members and total sales of more than £350,000 a year, of which the Coronation Street branch is an off-shoot. The results attained were traced to various causes: "the price and quality of the goods, bright attractive premises, small quantities sold, obliging shopmen, easy customs as regards entrance, dividend, etc., the support of committees and managers, and"—the key-note of the whole—"active propaganda." *

Thus Coronation Street has proved that it is possible for co-operation "to win the trade of the poorest," that such trade can be made a source of financial gain, and that "the store is a sure means for gaining the confidence and friendship of the people, establishing a relationship built on self-help, and unspoilt by the demoralizing effects of charity." It was further claimed that the store might

* *Women's Co-operative Guild Annual Report.*

become " a foundation and centre for constructive social work in parts where it is most needed."

One danger may perhaps be mentioned, arising from the simplicity and ease of the conditions of admission. The policy of "getting the people in," no matter how imperfectly equipped as co-operators they may be, has been frankly adopted, not only at Coronation Street, but in many other centres of co-operative activity. " We act," says the secretary of one society, "on the principle that it is the members and their trade that we want, and it is bad policy to place any obstacle in their way." * This is probably wise, or at any rate expedient, but it is worthy of note that the constant regrets that find expression among co-operators, that so many of their members care for nothing but the trading facilities offered by the stores and for the dividends, are often accompanied by opinions emphasizing the necessity of instructing members from the outset that co-operation ought to be understood to mean a great deal more than this. The easier the conditions

* *Report of the Women's Co-operative Guild Inquiry*, p. 9.

of admission to membership, and the greater the business attractiveness of the stores, the larger, other things being equal, will be the number joining, and the greater the risk that the proportion of what may be called the unconvinced members will increase.

Thus, it is clear that the greater the material attractiveness of the stores, the more necessary it is to bring other influences to bear which will remind new adherents that, in joining the store, they have also become members of what is traditionally intended to be something of the nature of a social and industrial fellowship. It is just this influence that the *personnel* of the experimental Sunderland branch was intended to exercise.

Whatever the dangers of prosperity and attractiveness may be, however, the basis of the store for almost every kind of social effort undertaken by co-operators remains an essentially sound and indispensable one. It enters naturally into the everyday life of the people, and the part that it, and indeed every reputable centre of retail supply plays, is one of essential importance. But the structure that should be raised upon the basis of the co-operative store is exceptional, and will always require architects and builders somewhat exceptionally qualified. It is upon the power of the movement at first to supply and gradually to train those so qualified that the extension of "poor stores" will, perhaps, depend, and the Women's Co-operative Guild could set before itself few more useful tasks than to arrange for the training of a staff the members of which would be fitted, not less by technical inside knowledge of the co-operative movement, than by sympathy to undertake such work, sometimes permanently in a single centre, and sometimes in carrying out a six or twelve months' special propaganda, on the general lines so hopefully begun in Sunderland.

It is a matter for regret and even for some surprise that where so much was being achieved, a time of trade depression should have been allowed to end, on the score of expense, the distinctive feature of the Coronation Street branch that gave it its exceptional value and interest—its lady resident workers. But this has happened, and on September 17, 1904, a resolution "to do without paid workers at Coronation Street" was carried by a large majority at a general meeting of the members of the Sunderland society. During the preceding half-year the society, as a whole, had been prejudicially affected by the depressed conditions of employment prevailing in the town, and in the declining trade that had resulted the Coronation Street branch had shared. The experience, however, of the whole two years from October, 1902, makes it at least doubtful whether even on purely business considerations an unwise step has not been taken in the apparent economy of saving an item of £150 a year. On the social side there has been no question of failure, and the whole discussion at the meeting turned on the question of expense.

It is, of course, not intended to discontinue the valuable work of the past, and it is hoped that it may be possible to carry it on by local voluntary service, especially of that of members of the Women's Co-operative Guild and of the Education Committee of the parent society. If these bodies realize the exceptional responsibilities which the withdrawal of the grant for the "Settlement" workers places upon their shoulders, no great harm may result so far as Coronation Street is concerned; but the co-operative movement, as a whole, has lost, at any rate for the moment, the stimulus of a significant and very promising departure. There is reason to hope, however, that neither the other societies * that were proposing to follow the Sunderland example, nor the Women's Guild itself, will be greatly deterred by the Sunderland recession.

The spread of the "poor store" is not, however, likely to be rapid, unless the flame of a new enthusiasm and of a new determination to realize old hopes burn into the hearts of large numbers of those who guide the movement both at headquarters and in the societies. The line of least resistance is not only giving them so much to do, as often to absorb most of their time and energy, but it leads them on, and for some time it will probably be pioneers alone who will tread, not where success is most assured, but where needs are greatest. The importance of the new departure is thus not to be measured by its achievements, considerable though they have been, but by the hopes and by the determination to which it may still give rise.

* At Hull, Stockton, and Sheffield. In a few other places alterations are being made in rules, etc., in order to facilitate the enrolment of a poorer class of member.

PART II—THE WORKSHOP

CHAPTER IX

CO-OPERATION AND "LABOUR"

The Position of "Labour"—Co-operative Achievements—Questions for the Future

THE distributive store, whatever its attendant social advantages may be, stands in the first place as the practical answer that has been given by large numbers of people to the question that they have either put to themselves or had forced upon them, as to how they could improve their position as "consumers." The more genuine members of the co-operative brotherhood are, it is true, apt to be critical of those who join their ranks simply because of the personal gain secured, but self-interest has nevertheless been the great underlying motive.

As a citizen a co-operator may gain, and it may even be argued that his position as a wage-earner is strengthened, by his increased independence, as a man, for instance, with some reserve of co-operative savings behind him. But this advantage is only indirect and incidental, even if it be secured at all. In joining his co-operative society and in determining its policy, it is not himself and others in their capacity of wage-earners that have been primarily considered. The questions of "labour" and the remuneration of labour are in this connection regarded as matters of secondary importance.

At a small centre of distributive dealing there is, indeed, not much room for the recognition of any such wider aims, although the question of the best form of business relationship is involved even in the engagement of a single employee at some little store. When, however, stores increase in size, the question begins to assume a new importance. When they undertake "productive" operations, this is still more the case, and when the action of the co-operative wholesale societies is taken into account, with their numerous productive works, a great question of principle can no longer be ignored. But there are very different ways in which it can be faced.

It is to the great credit of the co-operative movement that, in as far as it has been able to control matters, it has steadily set itself against all forms of "sweating." Commodities sold at working-class stores have generally been made under decent conditions. Those in the direct co-operative service rarely, if ever, habitually work for excessive hours; the weekly half-holiday at the distributive stores is almost universal, even in places, now rapidly diminishing in number, where the practice is not adopted by other traders; and wages are nearly always up to, and sometimes a little above, the current and accepted trade union or other local rate. It is in the salaried positions alone that remuneration is still, and has been far more often in the past, below average rates; but, speaking generally, the conditions of "labour" under the co-operative *régime* are, as compared with the outside prevailing standards, good.

Is this enough? Ought the co-operator to establish some generic difference between the position of those in co-operative service and those in receipt of a weekly wage in ordinary capitalistic enterprise? What ought to be the co-operative bond between "employer" and "employed"? Is there something anomalous in the employment of men by co-operative societies for a weekly wage, and in the neglect of any effort to establish some closer, more permanent, and more moral bond? If there be something anomalous, in what does it consist? Do we look for something else because true co-operation bids us? or because in the general interest we look for a new principle that shall make for increased efficiency? or because something more is required to make the position of the wage-earner more satisfactory and more progressive? Is our quest, if we undertake it, to be carried out in the interests primarily of the co-operative movement itself, or of employees as a class, or of the community?

To such comprehensive questions, no summary answers of general applicability to the varying conditions of trade and employment can be given, but before attempting to see what kinds of answers may be possible, it will be convenient to consider in somewhat greater detail what the practice of co-operators themselves is.

CHAPTER X

THE POSITION OF DISTRIBUTIVE EMPLOYEES

The Position of Distributive Employees—Local Differences of Policy

FIRST as to the distributive stores. It has been already stated that the usual practice among these is to engage the permanent employees at a specified salary or wage, but in 1905 228 societies in Great Britain, out of a total of 1452, slightly modified this relationship and allotted a share of profits to employees, as is shown by the table on p. 156.

The table shows that the practice is far more generally and somewhat more effectively adopted in Scotland than in any other part of the kingdom, the average amount allotted there amounting to about £4 3*s.* per employee in the year, equivalent to 1*s.* 7*d.* per week, as compared with about £2 4*s.* 4*d.* in England and Wales.

PROFIT-SHARING BY CO-OPERATIVE SOCIETIES

Returns of Amount and Percentage on Wages of Profit allotted to Distributive Employees by Retail Distributive Societies

Year	Country	Total number of societies	No. of societies making returns	No. of employees	Amount of wages paid	Allotted out of profits	
						Amount	Percentage on wages
					£	£	
1905	England and Wales ...	1148	185	10,619	558,705	23,578	4·2
	Scotland ...	282	36	3,403	154,799	14,129	9·1
	Ireland ...	22	7	110	4,777	271	5·7

From the "Eleventh Abstract of Labour Statistics, 1907"

Within the boundaries of England and Wales, however, the practice shows marked local variations, and over large areas is sometimes hardly represented at all. Thus, according to the 1906 report of the Co-operative Union, in its northern section, in which there are nearly 150 societies of every kind, paying about £480,000 annually in wages and salaries, only £366 were paid during 1905 in the shape of dividend to wages; while in the southern section, in which a somewhat smaller amount is paid in salaries and wages, this dividend amounted to more than £10,000.

Of the two Wholesale societies, in the

Scottish alone is the practice adopted, and in this society in 1905 £3,972 was allotted as special dividend to the 1634 employees of the various distributive departments, a sum that was equivalent to 3·3 per cent. on the total wages paid.

Profit-sharing—that first step towards an altered wage relationship—is thus seen to be a plant, not only of somewhat feeble, but of very uncertain growth in the distributive side of the co-operative movement.

Before giving particulars as to the wage relationship existing in co-operative productive undertakings, it will be desirable to give some figures showing their extent. In the following pages, co-operation as applied to agriculture is excluded. It raises many special questions, is for the most part different in aim and conception from that which has come to be known as ordinary co-operative enterprise, and will be dealt with in separate chapters.

CHAPTER XI

CO-OPERATIVE PRODUCTION

Co-operative Production—General Table

CO-OPERATIVE production exists at the present time in three fairly well-defined forms—in the factories and workshops of the two great Wholesale societies, in connection with the retail distributive stores or with federations of these, and in separate productive associations. Because of their extent and long history, rather than because of any distinctive characteristics, a fourth group is generally distinguished in the Corn Mill Societies.

The following table will give the essential particulars of all four groups :—

CO-OPERATIVE PRODUCTION

I. *Productive Branches of Retail Distributive Societies*

Year	No. of returns	Persons employed	Value of productions £
1895	494	8,854	2,356,405
1905	873	19,227	6,370,991

II. *Productive Branches of Wholesale Distributive Societies*

Year	No. of returns	Persons employed	Value of productions £
1895	2	6,684	1,570,598
1905	2	15,569	6,154,113

III. *Corn Milling Societies*

1895	9	404	957,908
1905	8	420	1,378,328

IV. *Other Productive Associations* *

1895	170	6,716	1,393,928
1905	166	8,877	1,847,320

* Excluding Irish dairies.

CHAPTER XII

RETAIL SOCIETIES AS "PRODUCERS"

Retail Societies as Producers—The Example at Rochdale—Guiding Principles—Departments

THE Rochdale Pioneers, as we have seen, desired much more than "the establishment of a store for the sale of provisions, etc.," but the lines of success guided their efforts primarily into this channel. Although productive enterprise has followed, the most salient fact in the history of the society has been its success as a centre of retail distribution, and what people now mean by the "Rochdale System" is connected with the particular features of its constitution as a distributive store. "The interests of the consumer" have thus come to be the accepted explanation of the dominant motive, not only of purely distributive societies founded on the Rochdale model, but of the productive enterprises in which they frequently engage, and the phrase fairly represents their point of view.

The starting of productive departments of distributive associations is the natural and economic sequel to the possession of a fairly assured market, and of the necessary capital. It is the modern practical business-like and somewhat restricted form that the hopes of many forerunners of co-operation took when, as at Rochdale itself, they hoped to "organize the power of production."

It is easy to imagine the position in a prosperous store possessing some thousands of members, with a steadily increasing supply of capital available, and crying out for fresh developments. Perhaps in the district in which such a store is located there has been a difficulty in getting good and wholesome bread. Whether this be so or not, the capital is available for arranging to supply a commodity that every member needs. The committee, urged on perhaps by the representations of the members at their periodical meetings, proceeds to make the necessary inquiries as to experience elsewhere, probable outlay involved, etc.,

M

and decides that it should be possible to make the project a valuable and paying department of the society. From first to last the proposal has been weighed with the general interest of the members of the society in view. The bakery is built and staffed; the highest local rates of wages are paid; considerate arrangements are made as regards the length of the working day; the conditions under which the men are to work are made thoroughly sanitary and satisfactory; and the society, we may assume, wishes in every respect to prove itself to be a "model employer." The employees probably are, and at any rate can be, members of the society, and can thus secure their share of the half-yearly dividend on their purchases, including that accruing from any profit that the bakery itself may make.

Such is very much the story of the starting of most of the productive branches of distributive stores, and such is as a rule the point at which their industrial task may, it is considered, rightly stop. In starting some fresh department, the unsatisfactory local conditions of the section of labour concerned may, it is true, have entered into the calculations of both the members and their committee, and the betterment of these conditions, in so far as these could be affected by those whom the society is proposing to employ, may have added something to the argument in favour of the new departure. But the effect upon "labour" is only very rarely the cause of the action taken, while the idea of placing the labour employed, in any condition other than that of wage-earners working under good but strictly normal conditions, is, as in the case of the distributive branches themselves, as a rule absent: * "the interest of the consumer" is the paramount consideration.

In the particular directions chosen for their productive departments by the stores, the same consideration for the consumer naturally takes the first place, and it is thus found that nearly all these departments are started for the manufacture or preparations of commodities in the most extensive and most ordinary demand among the members. Thus, in 1905, out of a total of 2016 of these productive works carried

* In 1905 returns were obtained from only 114 productive departments of distributive societies, employing 2465 persons, in which a bonus to labour was given. In these 114 cases a total of £7126 was paid as dividend on wages, representing in England and Wales 4·1 per cent. on earnings, and in Scotland 7·5 per cent.

on by 873 different societies, nearly 750 were engaged in food preparation, chiefly the making of bread, preserves, confectionery, and in corn and meal milling, while nearly 1100 were clothing factories of one kind or another.* Of the total sales for the year credited to these various departments, more than $4\frac{1}{2}$ millions out of a total of nearly $5\frac{1}{2}$ millions is accounted for by the food preparation group, and considerably over $1\frac{1}{4}$ millions by the clothing group, the latter also accounting for 11,335 out of a total of 19,227 employees. The only other considerable productive group in addition to the two mentioned is that of building, carried on in fifty-seven different societies, reflecting for the most part the attempts of co-operators to do something towards providing their members with what is in as general demand as good bread or good boots—namely, good houses—and accounting for 1672 employees, and sales to the amount of more than a quarter of a million. The sales under the head of farming, dairying, and fishing, spread over fifty-four different societies, amounted only

* Boots and shoes, 493 ; tailoring, mantles, shirt-making, etc., 298 ; dressmaking, millinery, etc., 296.

to £102,881.* Other industries were represented to a quite unimportant extent.

* Among other interesting developments that may lie hidden behind the recital of bare figures, the achievements of the society at Desborough—a place that is the centre of a variety of co-operative undertakings—deserve notice. The distributive society here not only affords one of the somewhat rare instances in which farming has been successfully undertaken on a fairly large scale, but the property of 408 acres of which the society is now the freeholder contains a valuable bed of ironstone, the working of which the society is keeping, and, so far successfully keeping, in its own hands. In this case, therefore, co-operation is breaking new and important ground. On the non-industrial side there is also a point worthy of note in connection with this society, inasmuch as the property owned carries with it the right of presentation to the living of the local parish church, a right that the members, most of whom are Nonconformists, have with great good taste handed over to the Bishop of the Diocese.

CHAPTER XIII

THE WHOLESALES AS "PRODUCERS"

Productive Departments of the Wholesale Societies—Policy as regards the Remuneration of Employees

WHEN we turn to the English Wholesale Society we find the same simplicity of productive aim: the making under satisfactory conditions * of commodities in everyday demand. The modest start made in 1873, when the Crumpsall Biscuit Works were commenced, has been followed by similar development in many other directions, and more than 10,000 employees are now engaged in the various productive departments of this great society, which, to quote its own enumeration, now manufactures "flour, butter, biscuits, sweets, preserves, pickles, candied peel, cocoa, chocolate, tobacco, cigars, cigarettes, snuff, soap, candles, glycerine,

* For instance, in more than half of the departments the working hours are under 48 per week, and in no case does the week exceed 55½ hours.

starch, boots and shoes, saddlery, woollens, clothing, flannels, shirts, mantles and under-clothing, corsets, millinery, hosiery, silesias, pants, ladies' underwear, cardigans, furniture, and brushes."

From 1895 to 1905 the value of the output of the various productive departments increased by about 326 per cent., in the latter year amounting to nearly 4 millions as compared with nearly a million in 1895.

In this society there is a consistent refusal to admit "labour" into any form of individual partnership, and thus to any share in "profits," except in so far as these may percolate through in the shape of dividend by transactions with the Wholesale, and paid by the constituent society of which the employee may be a member. In connection with the English Wholesale Society, the position of the employees is considered by co-operators to be of exceptional importance, because, within the borders of their movement, the productive operations of the English Wholesale are so extensive. They represent, with the exception of the similar departments of the Scottish Society, the most expansive side at the present time of co-operative production. When it is remembered, moreover, that in some cases these offshoots of the activity of the Wholesale come into active competition with the independent co-operative undertakings in which the remuneration of the operatives is determined in accordance with what is regarded by many as a better and a more equitable basis, it is not surprising that we can trace one of the chief divisions in the co-operative world to the different views held as to the policy that should be adopted by the great federal society on this fundamental question of "wages" and the status of the wage-earner.

In the Scottish Wholesale Society there has been an increase of nearly 196 per cent. during the ten years 1895–1905, the total having risen in the period from £684,000 to a figure that in 1905 exceeded 2 millions. The total number of employees in this year in the productive works of the Scottish Society, including 306 engaged in the building department, was 5060 out of the aggregate of 6635 employed.

In this Society a bonus to labour is allowed, a uniform rate based on the results of the aggregate working of the Society being given, both to "productive" and to "distributive" employees. From 1884 to 1892 the whole of the bonus earned was handed over to the men entitled to it, but since the latter year a wiser plan has been adopted, and one half of each worker's bonus, which has for the last few years amounted to 8*d.* in the pound on wages earned, is retained and put to his credit. The special fund thus formed bears interest at the rate of 4 per cent. per annum, and accumulations cannot be withdrawn, save with the consent of the committee, until the expiry of three months after the service of the Society has been left. Another way of strengthening attachment is found in the admission of individual employees as shareholders, up to a holding of £50. Interest is paid at the rate of 5 per cent. on the stock thus held, and for every 150 employee-shareholders one representative can be sent up to the quarterly meetings of the Society. In 1905, however, fifteen years after the introduction of the scheme, there were 533 shareholders, or only about 8 per cent. of those who might have qualified—a result that is in itself somewhat disheartening.

CHAPTER XIV

"CHEAPNESS"

The Co-operative Attitude towards Cheapness—Disorganized Buying

"THE position of labour" in the greater part of the "federated" field of industrial co-operation is thus seen to be very similar to that in which it is found in ordinary well and considerately managed centres of employment, and it is because "labour" is thus left by the great bulk of modern co-operative enterprise in an unchanged economic relationship that the lines of modern development are most often criticized. We are even told at times that co-operation has "failed," and more often that "the old ideals" have no weight. Ideals, however, always dwell in the realm of the unattainable, and the hyper-critical have to beware lest they undervalue what has been actually accomplished. Between the undue depreciation of adverse critics, and the rather fulsome and exaggerated self-gratulation of confident admirers, it is equally necessary to steer our course, and it is, perhaps, to be noted that those who magnify into a moral precept the obligation of "loyalty to the Store" in the mere sense of making all purchases there, or who think that they see in the network of federated societies and in the Wholesales the germs of a "co-operative commonwealth," are, probably, as responsible for adverse comment as are, for instance, jealous traders, since nothing weakens a position so quickly as the erection of a platform of somewhat irrelevant moral appeal, or the exaggerated estimate of what has been accomplished.

Over and above their great demonstration of a power that has been self-contained, self-generated, and that seeks the acquisition of no power and of no rights at the expense of others, the great achievement of the "store movement" has been to spread comfort, and to increase the sense of responsibility and the power to meet responsibility among large numbers of the people. It has therefore often helped to make life more enjoyable and the conditions of well-being more assured, and, like other democratic organizations, it has, as regards a minority, helped to widen out the range of individual interests, mainly in the direction of education and civic duty.

Nor can it be denied that although no large claim can be advanced on behalf of co-operation as to what it has accomplished or tried to accomplish, in improving the status of wage-earners, its appearance in the field of employment has been almost always advantageous. As we have seen, Co-operation has always set its face steadily against the worst forms of competitive pressure, and even when non-co-operative products are purchased by societies there has been a more general recognition than is found elsewhere of the fact that cheapness is not in itself a sufficient justification for dealing, even though "good value" seem to be given. The difficulty is notorious that confronts the considerate person when he endeavours to assure himself that what he buys, be it raw material or finished products, has not been produced under conditions that are incompatible with decent living. It is well known, for instance, that expensive commodities are not infrequently made in low-priced workrooms, and the channels of connection between comfortable, spacious, and carpeted show rooms, and stuffy and crowded workshops, are very hard to trace. But as a movement, co-operation has tried to overcome this difficulty. In general, it has stood for "fairness" in industrial life, and although those buying on behalf of co-operative societies have often, like the members themselves, evaded or ignored their obligations in this matter, the extent to which the dangers and evils of excessive cheapness and excessive competitive pressure has been recognized is very greatly to its credit.

It might appear that the co-operative attitude towards articles that are economically suspect would be normal in a movement controlled by a class of consumer that in its own industrial experience must be alive to the exceptional dangers of disorganization and excessive competition. But it is not so, and appeals have been made again and again to co-operators that they should free themselves from the temptation to demand goods that are condemned by their very cheapness. Thus, a circular recently issued to the retail societies by the London branch of the Co-operative Wholesale Society said

"We would ask you to do all in your power to get your members to pay a fair price for all furniture. It is no satisfaction to ourselves to send cheap made furniture away, nor to yourselves or customers to receive [it] . . . we wish it to be distinctly understood we cannot hold ourselves responsible for furniture at prices which barely pay for material."

Those, however, who are most subject to the risks of competition are often those who are most ready to take advantage of the residual output, and form the most eager body of purchasers of commodities, the very presence of which upon the market is a demonstration of conditions, lurking somewhere in the body industrial, that must be entirely harmful.

Men in this connection are rarely ready to apply to themselves the principle that they perhaps in general are quite willing to accept, and probably most of us have ignored it at times.* In many cases, moreover, the individual income does not seem to allow the margin that makes selection possible on any other basis than that of price. "It is 'cheap,' and I must have it," is what many are almost forced to say, whether it be unwholesome food or shoddy wearing apparel that they are buying, and it is among those whose economic freedom seems thus curtailed that perhaps the most important work of distributive co-operation lies in the future.

Meanwhile many of those in the somewhat higher ranks of industry do not exercise the greater freedom they possess through their greater margin of income, but in their capacity as consumers add to the burden of disorganization that already presses on those who belong to the more helpless sections of the community.

* "Yes," a jobbing bootmaker in Whitechapel once remarked to an inquirer, "conditions in my own trade are bad enough, but the tailoresses working round here are far worse off. My wife, who is a Bible reader, will tell you all about it." At this moment the wife came in, and bore out by many facts her husband's indictment. "I will just show you," she said, "a suit I bought the other day for my little boy, and that will give you an idea of what earnings must be." With the words she fetched a small suit, braided and ornate, for which she had paid about 7*s*. 6*d*. But she was quite unconscious of any inconsistency between her sympathies and her actions.

Disorganized buying by the comparatively well-to-do workingman's wife, often late at night, with cheap bargains as the sole consideration in her mind, is one of the most fruitful sources of disorganized selling and disorganized making. Quite apart, therefore, from any more ambitious aims in the solution of the "labour" problem, it is a great gain that the co-operative movement has in general steadfastly set its face against these more subtle and perhaps more widely spread sources of industrial evil.

If the stores do nothing more than spread the practice of orderly and thoughtful expenditure, and of the purchase of wholesome commodities, they will have more than justified their existence, and the community at large will owe them a great debt of gratitude. For perhaps the greatest social need of the present time is disciplined life, good home management and wise expenditure, and in securing all these the co-operative societies have helped not a little.

CHAPTER XV

INDUSTRIAL CO-PARTNERSHIP

The Status of the Wage-earner—The Trade Union—A Co-operative Workshop—Consumers' Associations and Productive Societies—Profit-sharing—The Productive Federation—Labour Dividend—Need of Variety in Business Structure

THE question of "labour" however remains, and to many ardent spirits, to-day as in the time of Robert Owen and of the early Christian Socialists, the "emancipation" of labour represents still by far the greatest objective of industrial reform. They look to the day when all men shall be recognized as partners in the vineyard, and are impatient of a relationship that still leaves the husbandman, no matter what the character of the lord of the vineyard may be, a hired servant. Those, therefore, who find the chief sign of progress in a change of status, and in a tendency to merge in one the well-defined functions of employer and employed, and who accepted it, moreover, as an essential part of the co-operative faith that this change must be always made a part of co-operative practice, are among the severest critics of what seems to be the dominant trend of the co-operative movement of to-day.

We are thus, as Mrs. Sidney Webb emphasizes,* confronted by two co-operative ideals—one of an association of consumers, with salaried or wage-earning officials, acting under the general direction of elected committees in the interests of the whole community; the other of groups of workers who, in virtue of their economic position, secure as individuals some special share of the profits that may be made, and, it may be, some voice in control. Under the first scheme no man secures a particularized individual profit; under the second, no man is simply a wage-earner. The association of consumers as a federalized movement grows from within, and if it embark on the field of "production," must find its market among its own members. The other group of workers may or may not find a market

* Cf. "Problems of Modern Industry," by Sidney and Beatrice Webb, pp. 193-194.

among co-operators, and although up to the present time it is to co-operative consumers that most of their products go, it is open to them to seek their outlet and their profit anywhere.

Although these two ideals are in theoretical opposition (the former, it may be noted, merging into the conception of the socialist state or "co-operative commonwealth"), and in practice sometimes conflict, the field of industry actually covered by co-operative enterprise of every kind is so small as compared with the whole, that there should be abundant room for both.

The association of consumers has, as has been seen, its own distinctive social task, and in some of its aspects no plan of financial partnership has much value, either economic or moral. Especially is this so on the more propagandist side of the movement, and in the extension of distributive co-operation, with its attendant social advantages, to those low down in the industrial scale. "Profit-sharing," for instance, perhaps would not "pay" here, and even if it did, it would often show that those who personally benefited by its adoption were not best fitted for the propagandist posts into which they had been put. This may be recognized; but, on the other hand, when we hear that the employees of some co-operative productive enterprise are in many cases not even members of the local co-operative store, and are hardly conscious that they are working in a shop the profits of which are distributed over a great multitude of working-class families by whom the capital is supplied, and who look upon their work as just the ordinary task that the wheel of life has brought them, and from which they would gladly be free, the feeling of regret cannot be suppressed that democratic industry has so completely failed to touch the imagination and to lift the life of those whose services it requisitions. Strong though this feeling may be, however, it is nevertheless, as has been already urged, more useful to concentrate attention upon the tasks that the association of consumers is achieving, and has it in its power to achieve, than to find fault, even though it be considered that some of the wider, if not the higher, aims of industrial reform have been put on one side.

The individual co-operator is, however, not simply a member of a distributive store. He, with millions still outside his ranks, desires to combine with comfort in his home a satisfactory position in the workshop, or factory, or yard, which in the vast majority of cases lie quite outside the scope of the co-operative movement itself. In securing this position he knows that his individual personal qualifications—skill, good temper, reliability, strength, etc.—are the chief determining conditions, but he seeks often to make his economic position still more assured, and in combination he has so far discovered his most distinctive instrument. The trade union has been invoked, and in many great industries it has achieved much in strengthening the position of the individual worker. But its usefulness and power are limited. In aiming at approximate uniformity of condition in any single trade, it often turns the balance too much in favour of the average man, and, whatever its effect upon aggregate efficiency, it tends to leave impersonal and formal that very relationship between employer and employed that should be intimate and friendly. Invaluable as an instrument in the process of industrial evolution, and still often indispensable, the modern trade union can hardly be said, in itself, to represent the most satisfactory and the most durable relationship.

Adequate capital, its capable use and the association of the appropriate labour, are the three strands out of which the cord of stable industry is spun, and it is essential that there be the fullest and most conscious recognition of this fact by all parties concerned.

Although, however, an underlying community of interest were recognized, there would remain the question of sharing the proceeds, and it is in the attempt to cut the knot of this problem of problems that the fully developed co-operative workshop, owned and managed by the "workers," has been often advocated. It has, however, rarely succeeded, the practical difficulty, even if the necessary capital has been forthcoming, of merging the functions of management and labour—of "master" and "men"—having generally proved insuperable. It is, however, in such a workshop—the capital owned and the manager appointed by the operatives working in it—that the practice of industrial co-partnership finds its completest although not its most hopeful or most successful illustration.

The commonest type of co-operative productive society in England, apart from those started by the consumers' associations (wholesale or retail), rarely rests upon the basis of individual membership, and finds its chief support in the distributive societies themselves, which, subscribing a portion of the capital and, with similar societies, being generally the chief customers of the productive society, naturally take an important share in its management.

Out of a total of 132 productive associations, the accounts of which were analyzed for the Board of Trade Report on Co-operative Societies, 44 per cent. of the total sales, that is, £696,436 out of £1,573,121, was attributable to twenty associations, in which the whole or the majority of the members of the management committee were representatives of retail distributive societies. Other facts illustrating the constitution of these productive associations may be given. Out of the total number of committee men engaged in managing these 132 associations, only about 30 per cent. were employees. Of the total individual membership in Great Britain only 15 per cent. consisted of employees, and by these only 9 per cent. of the total share and loan capital was held. In fifty-five of the associations no employees were on the management committees. Only in the twenty-five boot and shoe and tailoring societies, with an aggregate membership of the committees of management of 269, did more than half of these consist of employees.* For 1905 particulars were obtained from 389 societies (including 264 in Ireland), and in these, out of an aggregate of more than 9000 employees, as many as 60 per cent. were non-members. By the employees less than 6¼ per cent. of the total capital was held, and they made up only 6·8 per cent. of the various management committees.†

As regards ownership of capital, membership, and participation in management, therefore, the position of employees is seen to be a subordinate one in the very group of societies which, apart from the Irish societies included in the above figures for 1905, exist primarily

* *Report on Workmen's Co-operative Societies*, 1901, p. 55.

† "Eleventh Abstract of Labour Statistics."

in the interests of producers, and the facts cited are a reminder of a truth that demands far wider recognition than it yet receives that the formal constitution of the business unit throws but little light on its real character.

In the great majority of the associations to which the facts refer, the employees share in the profits by a constitutional right, and the acceptance of this principle, coupled with a fixed rate of interest on capital, is the cardinal feature of all acknowledged partnerships in industry, when capitalists, purchasers, and workers, all in virtue of the economic parts they play, and the last mentioned over and above their wages, share in the proceeds of the undertaking.

Among that section of co-operators themselves who attach primary importance to the idea of "co-partnership," the first place is given to the acceptance in some form or another of this principle of profit-sharing, not so much because of its own intrinsic value, as because it is the necessary stepping-stone towards promoting that feeling of responsibility and of interest that is calculated to lead to a more fully developed form of association. The

Co-operative Productive Federation, for instance, makes this test of profit-sharing a condition of admission to that body—

"No society can join the federation which does not by its rules practise profit-sharing with its workers, under which, in the first place, a substantial and known share of the profit of a business belongs to the workers in it, not by right of any shares they may hold or any other title, but simply by right of the labour they have contributed to make the profits." *

The second principle is that the worker invests his profits or other savings in shares of the society or company, and thus becomes a member "entitled to vote on the affairs of the body which employs him."

But in the federation itself only fifty-six societies † were enrolled at the end of 1905,

* "Co-operators' Year Book," 1907, p. 147.

† Comprising fifteen boot- and shoe-making societies; eleven formed in various branches of textiles and clothing; eight building and wood-working; nine printing; two watch-making; and one each for making padlocks, cutlery, needles, buckets and fenders, pianos, umbrellas and "co-operative sundries," besides engineering, barge-owning, book-binding, and type-writing societies. Seven of the societies were at Leicester; five at Kettering; four in London; and two each at Bradford, Derby, Coventry, and Wellingborough. The oldest and one of the best known is the Hebden Bridge Fustian Society, founded

showing a total trade for 1905 of £687,678, a profit of £23,768, and an aggregate dividend on wages of £3731. In the cases of eighteen societies no profit, and thus no dividend on wages, was shown.

From the Board of Trade returns for the same year* we are able to add on account of productive departments of retail distributive societies, and of the two large Scottish societies—the Wholesale and the Glasgow United Baking and other productive societies—which do not belong to the Productive Federation, about £24,500 to the profit divided, bringing the total dividend paid in these various societies in Great Britain up to something over £28,000. For 1903, the total for forty-seven societies making any returns under this head out of an aggregate of 126 societies, was put at £23,356, a sum that is nearly three times as great as that given for 1893.†

The total is, however, still small—so small,

in 1870. The large Leicester Hosiery Society no longer ranks as co-partnership, it having been acquired, in spite of some heartburning that the transaction caused, by the Co-operative Wholesale Society in 1903.

* "Eleventh Abstract of Labour Statistics."

† See *Labour Co-Partnership*, August, 1904, p. 123.

in spite of its steady growth, as to seem almost insignificant, and when the facts are recalled that as regards membership and control, the position of wage-earners is weak in most of the societies, we might seem to be driven to the conclusion, especially when it is remembered that considerably more than half of the above total dividend paid on wages is accounted for by the Scottish Wholesale and the Glasgow Baking Societies, that "co-operation" established primarily in the interests of producers has no special claim on our attention. It is clear in any case that it does not spread easily, and that something stands in the way of its rapid expansion.

Perhaps the most important conclusions that can be deduced from the modest position that productive co-operation at present occupies, is that a satisfactory and appropriate form of business structure must be contingent not only upon the particular characteristics of individual trades, but also upon the available reserve of business capacity and mutual trust. The whole canon of the co-operative faith may be true, on certain premises, but it is clear that in practice it can only be acted upon partially, as conditions may determine. It is a remarkable fact, for instance, that one of the most successful co-operative productive societies in existence, some consider almost the only genuinely independent democratic unit, was started when the operatives were exhausted by a strike and quite unable to raise any capital of their own. All of it, therefore, was loaned by outside sympathizers, and one of the first conditions of responsibility was thus apparently lacking, and thus also one of the first conditions of stability and success. But the right men were members. Above all, the right man was found to fill the position of manager, and the success that few ventured to anticipate, has been won. Even among co-operative undertakings, therefore, conditions will have to determine constitution, and those who boggle at variety and elasticity, either within or without the co-operative field, will be in danger of finding themselves in the grip of a stifling and restrictive orthodoxy.

CHAPTER XVI

THE LINES OF PROGRESS

Different Aims—Value of the Co-partnership Principle—Employers and Employed highly Complex Classes—Reasons for the Persistency of Status—The Organic Unity of Industrial Life—The South Metropolitan Gas Company a "Complete Co-operative Co-partnership"—Its Origin, Constitution, and Results—Conditions as determining Form—Mr. George Thomson and Woodhouse Mills—The Need of Eclecticism—Difficulties, Economic and Psychological--Note on "Gain-sharing" and *Les Sociétés Anonymes de Travail*

THE attitude of every progressive movement is one of anticipation, and those who look forward to better things are led to do so because certain conditions of the time seem to them to be either imperfect or positively undesirable. The socialist or the trade unionist, the educationalist, or the extreme temperance reformer, have each their special objective—based on their special antipathies or special hope, and although, when they come to close quarters with their more particular problems, they are apt to find that an apparent simplicity of aim disappears or is confused in the tangle of a highly complex relationship, some dominant hope and aim remain. The convinced socialist believes, for instance, in land nationalization, in spite of his discovery that much real property is mortgaged to friendly societies or building societies, with some millions of working-class members. The trade unionist believes in the standard minimum wage, although he sees that it may hasten the end of the industrial life or increase the difficulties of some of those in his own trade, who from age or some more than average limitation of powers cannot earn it. The educationalist is often undeterred in the pursuit of his more immediate aims by the exacting claims of health and health preservation; and the abolitionist is apt to regard the claims of the moderate drinker as of no account. Thus deadlocks may ensue, or absurd claims may be advanced, but progress nevertheless wins its way by this insistence of the major enthusiasms.

One of these which perhaps in industrial life is destined to prove of exceptional value in the future is this very ideal of co-partnership in industry, in spite of the fact that after some forty years of trial it still seems to have achieved so little. Alike on moral and economic grounds it is, however, full of hope. It frankly accepts the underlying solidarity of the interests of every industrial class, and it makes possible that variety of business constitution, which when necessary leaves management untrammelled. It is one of the forms of industry which is waiting for men as they travel on the upward road of foresight, restraint, and good-will.

It would be erroneous, however, to infer that the formal modifications of the position of wage-earner, which the more ardent school of co-operators desire, even if it were practicable, would be necessarily advantageous at the present time. Evidently it is not equally practicable, for the wage-relationship varies infinitely in character, from an almost momentary bargain to a life-long adherence; from the engagement of the unit, unknown and uncared for, picked out from a disorganized crowd of semi-skilled that can always furnish others if those selected fail, to that of the skilled member of some highly organized trade with every condition of his employment elaborately regulated by his trade society, and with a strong barrier erected against those without its pale.

On the side of the employer there are corresponding differences. His trade may be new and of uncertain future, or it may be well established and the records of a long experience may be known. The trade may lend itself easily to measurability of result, or the reverse. Responsibility may be highly concentrated or widely diffused, reaching not only to manager and foreman, but the rank and file of the employees.* Or, again, normal conditions may be almost common property, or any kind of publicity may be vigorously deprecated. The "employer," too, may be a Government dockyard, a municipality, a great joint-stock

* The principle of co-partnership is capable of an adoption that is more or less complete, and it is, in its most perfect form, also the most perfect realization of the principles of co-operative industry. It must be noted, however, that business arrangements in which special privileges or rights are allocated to foremen or heads of departments, but not to the rank and file of the wage-earners under them, often fall into another category. "Labour" here has no special recognition, no hand is held out to it, and while businesses so organized, which are very common, often help to ensure very great efficiency, they can hardly be said to involve the recognition of any co-operative principle, any more, for instance, than does the master builder who offers the inducement of a special wage to the leader whose business it is to set the pace for a group of operatives working under him

corporation, a millionaire, or a struggling beginner, who has, it may be, just left the ranks of the wage-earners themselves in order to start in a small way of business "on his own account." Finally, the employer may vary in temperament. He may be a hard man who clings to an obsolescent autocracy of relationship, and who feels that the claims of justice are sufficiently met when the conditions of the market, no matter how exacting and socially harmful they may be, are observed. Or he may be a man who sees through the circumstances of the moment, and is willing to give thought to plans by which, if he be met halfway, a genuine personal relationship may be established, lasting and mutually beneficial, making no less for good-will than for efficiency. Thus in every case that has been mentioned, in some way or another, the relationship between employer and employed, and thus the possibility of applying the principle of co-partnership, will necessarily be affected.

But just as the trade unionist or the socialist may have forced on his notice facts of life that conflict radically, and not simply through partisan or sectional differences, with his view of what industrial relationships ought to be, so does the advocate of co-partnership in industry see, if he endeavours to take a comprehensive survey, that the position of the simple wage-earner is often much more unassailable than that of the man who seems to have forced the barrier of division between "employer" and "employed."

In certainty of incomings and in freedom from responsibility, we are reminded, for instance, of some among the special advantages that the wage-earner frequently enjoys.

The fulcrum of his life is provided by the part that he plays in industry, but his working hours do not make up this life, and many wage-earners are to be envied who, a "fair" day's work done for a "fair" wage, are free in a very exceptional sense to use the leisure that their work secures. Moreover, just as there is a constant tendency to depreciate and lose sight of the importance of the individual industrial task, however humble it may be, and to forget that it is upon the aggregate industrial activity of the community that everything else depends—national safety and national education, art, drama, and literature, for instance, not less than the stability of some particular business form, or the support of some working-class home—so also is there the danger of over-estimating the power of material interests to determine action, especially when such action, although supposed to lead to reform, involves anything of the nature of sustained effort over and above that which the normal working day demands.

Apathy, although sometimes born of hopelessness, is probably more often a sign of contentment, and of a life that is finding, it may be in the home, in games, in amusement, occasionally in religious association, more often in the public-house—in the hundred and one ways, good, bad, and indifferent, that absorb the leisure of the working classes—a sufficiency of interest that leaves no margin for effort that is directed immediately to secure any alteration in industrial position. Happily, in spite of the immense obstacles placed in the way of industrial reform by betting, by extravagant (rather than excessive) drinking expenditure, and by vice, yet, as regards great numbers of people, there is in their accepted fixity of status little matter for regret. They are "contented," and their contentment rests upon an assured and reasonable economic and moral basis.

While the functions of the employer, however, remain as distinctive and as important as they are to-day, it is probable that over great ranges of industry the wage-earner will remain simply such, not only when he is contented and strong, either in organization or in personal qualities of skill or trustworthiness, but also when he is disorganized, weak, and degenerate. It is, indeed, at the foot of the ladder that men jostle each other most, and what seems to be the necessary retention of free initiative for those who hold the most responsible positions in industry and commerce is too often apt to be accompanied by an extension of the area of struggle. It is an old story, that for the motive of individual interest, which may mean, in practice, something very splendid or very sordid, no adequate alternative has yet been found; and any speedy substitution, therefore, were this possible, of industrial forms that may seem to stand on a higher ethical platform might only, in the present stage of development, substitute for the frequent suffering and deterioration of individuals that all deplore, the risks of national bankruptcy and, for great masses of the people, of starvation.

Industrial reform, therefore, can only be assured when its conditions do not conflict with still more exacting needs. Situated as we are, an imperial people with the commerce of an empire and of the world entering as an all-important factor in the life of each one, poor no less than rich, we are hampered by the burden of the very conditions that make us wealthy, and even projects for industrial betterment, if they weaken the springs of energy among those whose energy means most, must miss their mark, leading to loss instead of gain. Thus must all advocacy, all movement, be conditioned by the necessity of safeguarding the measure of freedom which is essential to that section of the community upon which the guidance and the expansion of industry most depend. The better "form" of industry will be useless, no matter what advantages it may seem to bring—even if it seems more just—if men are not ready for it, and it be found to diminish the initiative, energy, business capacity and width of view and aim among those by whom, in the interests of the community at large, it is most essential that these qualities be manifested.

It follows that employers, like Barkis, must be "willing," and must enter into this new contract with labour seeing its latent advantages, and believing that by the new sense of the community of interest that it is calculated to create, not only will the conditions of industrial peace be more assured, but that nowhere need the springs of action be sapped. "The first and chief difficulty," writes Sir George Livesey, "is that employers generally do not see its importance, and are not therefore disposed to take it up, and give the necessary time and attention to secure success; for it certainly will not work itself. It must be worked as earnestly and intelligently as any other branch of their business."* The completer the form of partnership adopted, the truer this statement is, and it is a very complete form that Sir George Livesey has in his own mind. Industrial partnership is

"much more than simple profit-sharing, which is but the first step in that direction. In-

* *Methods of Social Advance*, edited by C. H. Loch (1904), p. 112.

dustrial partnership, as I understand it in its complete and ultimate form, is a real partnership—beyond mere salaries and wages—of capital and labour, of employer and employed, in the business in which both are engaged; a partnership in capital, and therefore in the profits on capital, in responsibility, and in actual management." *

The successful experiment of the South Metropolitan Gas Company, with which Sir George Livesey is identified, affords in itself striking practical proof of the applicability of the principles which he advocates, and their original adoption as a counter-move to trade unionism is being forgotten in the spectacle of its beneficent achievements.

The place that the South London example is taking in the minds of many of those who are most keen for industrial and social progress, is reflected by the references made to it by Mr. E. O. Greening, a co-operator of nearly fifty years' standing, in his inaugural address delivered to the Co-operative Congress of 1904. Mr. Greening claims that the South Metropolitan Gas Company now

* *Methods of Social Advance*, p. 108.

constitutes a "complete co-operative co-partnership" in which "capital, custom, and labour are united by a bond of common interest." It is true that the participation of custom is determined in one essential feature by Act of Parliament, according to which the dividend on capital can only be increased as the company gives a drawback to its customers in the shape of reduced charges for gas on the price originally fixed, which forms, for customers and shareholders alike, the basis of calculation in determining the movements of their sliding scale of interest. This legal arrangement applies to other companies besides the South Metropolitan, but it is in this company that, apparently supported by other features more peculiar to itself, it has worked to most effect. The legal partnership was, however, incomplete, since the employees were not participants, and it was in order to widen the basis of partnership and to strengthen the bonds between the employees and the company, that the arrangement already legally adopted as between shareholders and customers was voluntarily applied to the workmen.

By this arrangement an annual bonus of

three-quarters or 15*s.* per cent. is given by the directors on the salaries and wages of all the officers and workmen, who enter into agreements with the company for various periods not exceeding twelve months.

The importance of the agreements is frankly admitted.

"If a profit-sharing bonus is given indiscriminately to all workmen, good, bad, and indifferent, it may very soon lose its effect as a stimulant to good work. Workmen who are careless and indifferent about their work are therefore told, when their agreement expires, that it will not be renewed until they show more interest in their work, and that they can apply again in three months. If they have improved, an agreement for perhaps three months is given, to be renewed if they continue to work satisfactorily. This system of agreements giving security for twelve months' work is greatly valued by the regular workmen, while those employed in the winter only have agreements for shorter terms, thus bringing them also into the partnership. The granting or refusing agreements is not left to any subordinate official or foreman, but all must come before the chief." *

* *Methods of Social Advance*, p. 115.

The power of discrimination, although essential to the success of the South Metropolitan scheme, of course differentiates it from the more democratic forms of labour co-partnership, but, the industrial qualities of men differing as they do, its reasonableness cannot be denied. It is one of those practical modifications of a more ideal relationship made necessary by a sane judgment and correct reading of the psychological conditions of the labour problem as presented by a great London gas company.

The success of the conditions is, however, well known, the bonus payable to those who entered into agreements in 1903 having amounted to £25,660.

The partnership of labour does not, however, end here. Rather it is the beginning, for the earning of a bonus which is simply distributed, and is not made the basis of any stronger tie, would leave the operative simply in the position of the man who, in certain conditions of the market when the gas was cheap, was able to earn something over and above his ordinary wages. His work might, and probably would, be better under such an arrangement, but the bond of connection would begin and end with the elasticity of his earnings.

Three plans are adopted to make the bond still closer and the interests more real and lasting: (1) Only one half of the bonus is made payable in cash, the other half being invested in the company's ordinary stock at the market price of the day. (2) The men have the option of leaving the withdrawable part of their bonus in the company's hands, when either 4 or 3 per cent. interest is paid according to the amount, or, if preferred, additional stock is bought. No less than 80 per cent. of the bonus is thus saved, and in the voluntary habit that has been thus formed is to be found perhaps one of the most conclusive proofs of the satisfactory character of the present system of management. It is a demonstration, not only of a social habit that is in itself of great value, but also of an attitude of mind towards the company that could never have been so fully proved by compulsory provisions no matter how wisely conceived these might have been. (3) Since 1898 two workmen directors have been elected by the shareholding workmen, these directors sharing equally with the others in the control of the company. The significance of this last arrangement is to be found, not in the voting power that it gives, but in the representative voice that it secures, and in its recognition of a new constitutional right.

Since co-partnership was introduced in 1889, nearly a quarter of a million pounds has been credited to the workmen, or about £50 per man of the large numbers employed, and more than 4000 men are shareholders. It must be noted that the period has not been free from the difficulties of fluctuating prices, the price of coal having sent up that of gas in 1892 to a level that brought the bonus on wages down from 5 to 3 per cent., and in 1900 to one that brought it down from 9 per cent. to zero. Success has therefore not been unconnected with very considerable difficulties.

As the result of the various advantages of the new order of things—material, moral, and constitutional—introduced into this company, it has resulted, writes Sir George Livesey, that "since 1889 a dispute, or a difficulty, or a difference of any kind between the workmen and the company has been unknown, and that such a spirit of co-operative good-will should prevail on the part of the men generally as to entirely relieve all that have to do with them from every trace of anxiety."*

At the moment the South Metropolitan Gas Company is the most conspicuous illustration of the successful adoption in capitalistic undertakings of the principle of co-partnership that this country can show, and its example is probably destined to be followed in many directions. At least four other gas companies, two in London and two in the provinces, have already adopted the same principle, and, as far as tested, with similar success.

In spite, however, of such examples, fortified by such a notable instance as that of the enterprise of the late M. Godin at Guise, the Labour Department of the Board of Trade knows of only sixty-two firms in the whole of the United Kingdom as having "adopted profit-sharing." †

In this, however, there is little reason either for surprise or for those that believe in the principle of co-partnership for despondency.

* *Op. cit.*, p. 108.

† See *Labour Gazette*, Sept., 1905, p. 263.

Just as it is "only a simpleton" who thinks that every good man must be cast in the same mould, so is it futile to expect any uniformity in business structure. For this, "goodness," as we have suggested, is conditioned by very varying circumstances of time, place, trade, and temperament, and even when there is a consensus as to what the most fundamentally important characteristics of satisfactory industry are—such as the development of a sense of individual responsibility, the maintenance of industrial peace, the economy of effort, efficient work, or, soaring to higher aims, the spread of the spirit of industrial fellowship, it is evident that the ways in which these characteristics can be developed, will not be the same, either in every trade or in every unit in the same trade—much less in every age. The industrial, like other forms of life, is feeling its way to the light, and in many directions as yet deep shadows fall.

In such a case as that of the South Metropolitan Gas Company the aim is comprehensive, and every non-political element of the industrial problem is touched. Even more radical was the moral and economic reform aimed at in the changes introduced by Mr. George Thomson at Woodhouse Mills, Huddersfield, in 1886. In this case, by a purely voluntary act and with the primary object of strengthening and placing on a firmer basis the spirit of "trust and confidence between men and men," an existing business was deliberately converted into an industrial partnership, almost the only limitation to a purely democratic foundation being found in the rule by which, subject to removal by a vote of five-sixths of the members of the association, the founder was made general manager "till his death or previous resignation." The step taken at Woodhouse Mills was that of an idealist, largely inspired by the teaching of Ruskin, who desired to apply in industry in the fullest possible way the great ethical conceptions of "truth, purity, honesty, and justice." Ideal though the aim was, success has been obtained, particularly as Sir George Livesey himself has recently written, "in the true spirit of co-operation which has been manifested by the workers."

As a rule, the degree of completeness in reform aimed at at Woodhouse Mills, or even by the directorate of the South Metropolitan

Gas Company, will be impracticable. Those responsible for new departures will be, and often will have to be, eclectic in their methods—an eclecticism that will be sometimes conditioned by their own ability, courage, or confidence, and sometimes by the intelligence and readiness to respond shown by those with whom they will have to act.

On the side of the employees, conflicting aims, such as those of the extremer forms of socialism, or the narrower and more exclusive and less progressive type of trade unionism, will sometimes block the way; sometimes it will be blank ignorance.

On the side of the employers, reform may be retarded by fear, lest change be followed, it may be, by undue interference with their freedom of action, or by what is regarded as a damaging publicity of trading results. The second danger, when it is one in reality, can often be guarded against by the accredited methods of accountancy, and the certification of results without unnecessary disclosures; the former, by the adoption of tentative and well-adapted changes in transition.

In practice, apart from the disturbing effects

of fluctuations in prices, of the uncertainty of profit and their temporary disappearance or conversion into a negative quantity, and of the mobility of labour, the normal hesitancy to entertain new ideas and adopt new practices often blocks the way. Just as a widespread conservatism often hinders the adoption of new methods as regards processes of manufacture, so does it also in the adoption of new methods of management, and still more of new methods of remuneration.

No great or rapid change in business structure is thus probable, and one great obstacle is found in the average shortness of the life of the business unit; but where there is prolonged stability, and especially where the joint-stock principle is adopted, future years are probably destined to see a considerable extension of the principle of co-partnership.

Changes, however, as they come, are likely, in the future as in the past, to present themselves in a multitude of forms. Sometimes a legal right will be conferred by the terms of a registered and formal deed of partnership. Sometimes there will be nothing but some participation undertaken on the part of employers as a voluntary act, to which custom alone gives any assurance of permanency. Sometimes only those will be recognized as participants in any new scheme that may be adopted who are stockholders as well as workers, and sometimes "labour" alone will be held to create this right. The proportion of the wage-earners' share of profits, and the method by which this is determined, will inevitably vary, as also will the method of government or representation on the governing body, and the period at which and the ways in which dividends will be distributed, and the various attendant conditions—as, for instance, as to whether the basis of sharing be the corporate or departmental results, or on the work of the individual participant.

In every case such variations will be the result of the special contingent conditions—psychological or industrial—and although some of these will be signs of hesitancy or fear, on the whole they will probably have to be accepted, just as those in the past have been, as genuine attempts to introduce changes that seem most suitable or practicable. It is noteworthy that Mr. Schloss, in his report on profit-sharing, made to the Board of Trade in 1894, draws special attention to the fact that in the various instances enumerated, "in hardly two cases is the system adopted identical in all particulars"; and although, as the principle becomes less tentative and experimental, certain more elementary conditions will be discovered which will be of more general applicability, a widespread diversity of detail will almost inevitably continue.*

* Differentiated from all forms of co-partnership is the method of remuneration to which the designation "gain-sharing" has been given. By this plan groups of workers, assured of a fixed minimum, are remunerated on a basis calculated on an accepted standard of cost, any extra earnings being, so far as the particular bargain is concerned, irrespective of any ultimate profits, or no profits, that it may give. If difficulties on the side of employers are overcome, such as the willingness to take the necessary extra trouble in organizing the giving out of work, or to incur the cost of bookkeeping on a more elaborate scale, an exceptional degree of cohesion on the part of the men will always be required on this plan of working, and this cohesion is unlikely to be found unless the number of those who share together be rather small and their individual efficiency be fairly uniform.

It is probable that, on the side of the men, somewhat analogous difficulties would attach to M. Guyot's proposed revival and extension of the proposed *Sociétés anonymes de travail*, as advocated in his recent publication, "Les Conflits du Travail et leur Solution." By these societies it is suggested that "labour" should be commercialized, the society undertaking under legal obligation to furnish at an agreed price products that may be required. The employer bargains, not with individual operatives, but *en gros* with the society. The terms agreed upon, the responsibility of the employers, save in paying, is shifted to other shoulders. With the workers, as such, he has nothing to do, even though they work in his shops. It is their product that is required, and if by the bargain this can be assured to time and of a quality as agreed, it is enough. The bargain is a commercial transaction. Certain products are bought, labour intervening.

M. Guyot starts with the assumption that the day of the paternal employer is not over, but doomed, and, holding that Trade Unions, Conciliation Boards, Industrial Arbitration, Co-operative Production, and Profit-sharing are all failing or have failed, either through the theoretical weakness of their fundamental position or through their proved powerlessness, he finds in these societies a machinery by which, while the normal difficulties that come from the old relationship between employers and employed may be done away with, a new influence will become effective, by which, since profit for all will be the real object in view, every physical and intellectual faculty will be brought into play.

Traditional custom, the limited powers of intelligence, combination and cohesion, the varying degrees of applicability to the new plan of different employments and different productive processes, will be among the influences that will probably make the growth of the plan advocated slow, although, like the practice of gain-sharing, it is already not without many illustrations. Granted circumstances that make an impersonal relation between employers and employed inevitable, or undesirable, and given the organizing and moral and mental powers required, commercialized labour would make for efficiency, and many of the present elements that lead to friction and waste ot time would be eliminated. As one of the methods of reform, however, rather than as the great solution, it is perhaps destined to take its place.

In form there is, indeed, no "solution" of the grave difficulties that the relations of employers and employed present. Many contributing influences will have to work towards reform, many of them permanently, since those, in whatever capacity, who take part in lessening the pressure of the industrial problem will perhaps never advance in single lines, or in massed battalions. Progress will always be *en échelon*, the units zigzagging forward, one company behind the other, the companies of very different sizes, constitution, and even aim. In the advance probably every method that has been mentioned will play its part and others.

CHAPTER XVII

OBSTACLES

Some Objections to Co-partnership and Proft-sharing—Opposed to Socialism—Relations to Trade Unionism

SOME considerable space has been given to the examination of the applicability, outside the ordinary working-class co-operative movement, of what are in reality co-operative principles, partly because of the comparatively slight extent to which any form of industrial co-partnership has been adopted by the co-operators of to-day, and partly because of the comparatively restricted field of co-operative employment, and because, on all the evidence and in view of the exigencies and pressure of industrial life, of its necessarily slow growth. It is in the great field of non-co-operative "capitalistic" industry that by far the greatest practical importance of the principle of co-partnership lies, if only because it is here that by far the greatest amount of the national capital is in use, and because it is, therefore, also here that the functions of employer and employed alike find their fullest scope.

But, as has been seen, objections may be raised to the principle of co-partnership itself, and especially to the practice of profit-sharing, which is in reality but the thin end of the wedge of co-operative industry. It is complained that the principle panders to the selfishness of men, that it would lead to the competition of groups of wage-earners, and that, destroying the feeling of working-class solidarity, it will be inimical to the trade-union movement. It is also felt, and felt more truly, that in as far as it led to a more diffused ownership of wealth, co-partnership lies athwart the path of socialism.

As to the last, the two ideals are indeed in fundamental opposition, if by socialism we mean, not only such extension of the function of government (local and central) as is being witnessed from year to year in this country, but the advocacy of a fundamental change in the economic basis of society. The useful consideration of practical economic reforms, however, has to start with the accepted basis of the private ownership of the great bulk of the capital, and the private direction of the greater part of the industry of the country.

As regards trade unionism, the difficulty and the antagonism would, it is true, arise in the case of industries in which the relations of employers and employed are determined by the action of the less scrupulous and less far-sighted class of employers. If these form the majority, or even an influential minority—and their power is often unhappily disproportionate to their numbers—the maintenance of a trade union on the present protective lines would be essential. But the moral atmosphere of many trades is changing. The more highly organized a trade becomes, moreover, the more do the functions of the trade unions become consultative, diplomatic, and pacific, rather than militant, and in the evolution of this change, rather than in the direction of their abandonment or weakening, the introduction of the principle of co-partnership would assist. It is interesting to note that so fervent an apostle of what he considered "the master thoughts of our own time, association, and service," as the late Bishop Westcott, should have seen even in the rival organizations of employers and employed "a step towards one organization of both." * At any rate, the basis of an assumed antagonism of interest between employers and employed can in no case afford a permanent starting-point in the organization of the personal forces of industry.

In the report on profit-sharing prepared by Mr. Schloss for the Board of Trade, the principal objections of the trade unions are summarized as follows:—

"It is not considered right that employees should have to take the word of their employer without means of verification as to the rate of profit earned. It is thought wrong that workmen who are discharged by or voluntarily leave the firm should lose their right to bonus for the current year, or should forfeit accumulation of deferred bonus. Arrangements under which participation of profits is confined to a certain number of the employees are disliked. Profit-sharing is objected to in cases in which the firm is considered to pay wages lower than those recognized by the trade union as the proper

* "Address at the Middlesboro' Congress Exhibition, 1901," p. 12.

rates; and it is sometimes considered that the existence of the profit-sharing scheme has of itself a tendency to keep the wages of the employees below the proper level. It is also thought that the adoption of profit-sharing leads to an undue amount of overtime being worked, and to work being done at unduly high pressure and with too small a staff, the services of extra workmen, who, as it is contended, ought to be employed, not being called into requisition. At the same time, the objection is taken that a particular profit-sharing scheme, or that the profit-sharing system generally, has the effect of depriving the employees of their independence, and making them unduly subservient to their employers. Where profit-sharing is believed to have been introduced with the object of weakening the influence of a trade union, the scheme is looked upon with very strong disfavour, while the employers who introduce profit-sharing usually, if not invariably, do so with this object in view, is an opinion sometimes entertained."

It is often thought that it induces—

"workmen to connive at the breach of trade-union rules, and to tend all round to undermine the power of trade-union combination, and" (as it is put in one case) "to remove the scope and field of operation of trade unions."

Many of these objections can, as Mr. Schloss points out, be to a great extent avoided by the adoption of the necessary safe-guarding conditions. Most of them, it is evident, however, are based on distrust, and on the feeling that one side is in some way or other sure to try and overreach the other.

The most vital economic solidarity, however, is not between "labour" on the one side and capital on the other, but between those of every grade who are banded together in a common industrial task. Combined with the maintenance of "fair," "live-and-let-living" relationship between the various units, the greatest industrial gain that the community could secure would be the recognition, subject always to conditions that make for national strength and well being,* of this internal community of aim and interest.

* The opinion is so widely spread that "the true commercial spirit cares little for the destiny or usefulness of the commodity it has produced when the sale is once made," to quote a sentence from Mr. Hobhouse's work on "The Labour Movement," that it is perhaps permissible to emphasize the importance of this qualifying clause. If it be not observed, as, for instance, by those who make or sell bad beer or unwholesome or adulterated food, then, no matter how liberally the scale of wages paid may be, how excellent the tone prevailing in the brewery or workshop or warehouse, or how perfect the formal relationship adopted

The greater part of the problem of applying modifications of the co-operative principle may indeed be said to resolve itself almost entirely into a simple economic and simple moral question—the one of gain, and the other of good-will.

The business life of the community is so complex—banking, credit, international trade, transport, production, distribution, consumption—make up in the aggregate so complex and bewildering a tangle of economic processes, that they are apt to obscure the fact that almost every bargain that these processes involve may be resolved into a series of personal dealings between men and men that are most satisfactory and most advantageous when intelligence and confidence are shared on both sides. Ignorance and suspicion are always calculated either to stop business or to make bad business. Under

between employers and employed, it is evident that admirable features of this kind cannot condone the wrong that is done to the community. The same sentiment that leads to an outburst of indignation should a War Office contractor try to defraud the nation by the quality of the food or clothing sent out to soldiers on active service is destined to find wider applications, and perhaps in the not distant future will bear with equal force upon any who play the same evil turn on the women and children and men who are engaged in the equally honourable service of the home and of civil industry.

normal conditions, therefore, and in every phase of business life that is not disreputable or fraudulent, the chief guarantee of economic worth and efficiency is found in the presence of the simplest intellectual and moral attributes. The former of these may vary both in kind and degree, and even as regards moral qualities, different occupations and callings sometimes give prominence to and make a special demand upon particular virtues; but as regards the mutual relationship of contracting parties, the moral attribute of good-will based on a confidence that is deserved on both sides is the most common necessity, and, of all that can be mentioned, the most potent safeguard of industrial peace and industrial efficiency.

This simplicity of the basis of industrial relationships is recognized by those who get through the tangle of the mass, and its importance, it may be noted, is just as great when there is collective dealing on one or both sides as when two individuals alone are concerned.

CHAPTER XVIII

A CO-OPERATIVE OPINION

Report of Co-operative Committee—Advantages of Profit-sharing — Creation of Privileged Class as Example to Employers

IT is, perhaps, the extreme merit of the advocacy of co-partnership in industry that it gives due recognition to the importance of sentiment in industrial life.

"The chief advantage of profit-sharing," states a report * made a few years ago by an influential committee † of the Scottish section of the Co-operative Union, "lies in its tendency to bring about a more harmonious relationship between employers and employed. When a satisfactory method of sharing is arranged, it is reasonable to expect that a better social spirit

* Reprinted in the *Scottish Co-operator*, April 29, 1904.

† The chairman of this committee was the president of the Scottish Co-operative Wholesale Society, and his colleagues were also well-known members of the movement.

will grow up between both parties; it will come to be realized that unity of interest makes for industrial peace, and so a better understanding and a kindlier feeling will prevail.

"Co-operators," the report continues, "are perhaps unconsciously drifting into the habit of looking upon the consumer as the only factor worth considering, forgetting that the only difference that exists between the co-operative and non-co-operative employer with regard to division of profit is that the one divides profits in proportion to purchases and the other in proportion to capital, and with few exceptions both alike decline to admit the worker as a factor equal in proportion and as worthy of consideration as either the employer or the consumer."

The committee was thus led to adopt the note, if not of criticism of the dominant policy of the movement to which its members belong, at least of appeal, and attention is directed to the important part that co-operators can play as leaders in industrial reform and to the weakness of their position if they hang back.

"The members of co-operative societies are mostly workers, but in their corporate capacity they are also employers. If, however, as workers, they believe the present wages system requires improvement, and wish to approach their employers on the matter, their position would be very much strengthened if they could point to those employed in the co-operative societies and say, 'See how we treat our servants; do unto us as we have done to them.' This would be setting an example, and the example would be valuable as an illustration of the advantage to be gained by the general adoption of the principle of profit sharing."

The economic as well as the moral advantages of the system are urged.

"It is contended that by this system sympathy and interest with the workers is aroused. He finds himself no longer a contributor to the wealth of his employer without hope of social advancement, and without incentive to personal effort. The knowledge that he will secure that share in the profits resulting from his labour is a strong inducement to careful and expeditious work, and prompts him to exercise a vigilance over his conduct which can hardly be secured even by the most rigid supervision under the eyes of an employer. He feels that he has a stake in the business, and that anything which tends to waste or extravagance is a direct injury to his personal interests. . . . The mutual sympathy and interest which is established between employer and workmen is calculated to minimize or entirely remove the friction and antagonism which too often exists under the present system of wage service."

This argument, in as far as it assumes that the quality of labour can be improved by the new system, is one that is often bitterly repudiated as appearing to cast a slur upon the integrity of the wage-earner. Experience, however, seems to show that it is not unreasonable to hold the opinion that, without any excessive and harmful exertion either of mind or body, many wage-earners could, beneficially to all concerned, work with somewhat greater intensity and with a greater sense of responsibility than at present. The advantages of the system advocated, however, are not so likely to be found in any alteration in the rapidity or degree of zeal with which some individual task may be completed,—the eager bonus hunter is no more attractive a figure than that of the piece-worker over-straining at his task—as in an altered attitude of mind towards, not only the task of the moment, but those that are to follow and those that others are executing.

Thus, forms of partnership that are of the most essential value, and that, in as far as they are formulated, should perhaps be regarded rather as the definition of obligations on both sides than as the assertion of sectional rights, ought to achieve something that is of far greater importance than securing an increment to wages: they ought to strengthen the possessory sense of the wage-earning classes, and thus deepen the feeling that they, no less than others, have a real and direct interest in the general prosperity. It is, moreover, in the spread of this feeling, and in the fuller realization of the economic truth that underlies it, by a steadily increasing number of the people, that the community can secure its surest safeguards in the future against political discontent, and its surest guarantee of orderly, social and industrial development.

To return to the report.

The objection is frankly met that since as a rule non-co-operative workers do not share in profits, to confer this right on co-operative workers would "create a privileged class."

"It may be granted," the report continues, "that we are trying to create a privileged class of workers, but the object in view is that all workers may become so privileged. All workers should share in the surplus or profit, and when once the method of dealing with our own workers has been put right, our aim lies in two directions—the first being to widen the scope of co-operative work, so as to embrace the largest possible number of workers; the second is to persuade all employers of labour to adopt our principle."

Alike, therefore, as an example in industry, as tending to secure economic gain and an improvement in personal relationships, and as a recognition of a more equitable method of remuneration, the principle was strongly advocated. "Co-operators should stand in the forefront," it is claimed, "of those who are striving after social reform of any kind, and we know of no subject more deserving of consideration than the improvement of the workers' condition." The further statement that "the present relation between employer and employed is out of harmony with the moral spirit of social progress," may overlook the facts already mentioned that the wage relationship is often far more satisfactory than any practicable substitutes, and that the trend of modern industry is often to differentiate even more sharply than in the past the functions of the employer from those of any other class—be it capitalist or wage-earner. But for many years to come the path of qualitative progress and of assured social and industrial peace, appears to lead so surely in the direction of well-devised and well-adapted schemes, based on this new method, not simply of the mere sharing of profits, but of the introduction of the basis of a more real partnership in industry, that there is little wonder that those who see its great social merits should be ready to believe that it will be followed more quickly and more generally than is likely to be the case. In all except its quite rudimentary and autocratic forms, the system is moreover well calculated to strike the imagination of men, to give a new sense of hope and a new spring to energy. Thus, many generous souls will doubtless strive towards it even though the road be often hopelessly blocked, and though there be still many failures to record to ensure the requisite combination of economic strength with social gain.

CHAPTER XIX

CO-OPERATORS AND HOUSING

Early Ideals—Investment of Surplus Capital in House Property—Development of Co-partnership Tenant Societies

ALTHOUGH in the early part of last century Robert Owen gave community dwellings an important place in his various schemes of co-operative life, and later again, in 1844, the Rochdale Pioneers, in their famous programme, laid down "the building, purchasing, or erecting a number of houses in which those members desiring to assist each other in improving their domestic and social condition may reside," as a prominent ideal, co-operators have until recent years made little effort to definitely apply their principles to the problem of providing dwellings for themselves.

It is true that a considerable amount of capital has been invested by societies in house property, returns recently made to the Co-operative Union showing that no less than

£9,603,438 had been so invested to the end of 1906. Of this, however, £6,532,296 had been advanced on mortgage of 32,600 houses to members wishing to build or buy houses for themselves; £1,232,073 had been expended by societies on houses which they have sold to members; while a sum of £1,839,069 only had been expended in building or purchasing houses which remain the property of the societies, and are let at rental to members. But there is no doubt that practically the whole of this nine and a half millions has been used in this way, not as a conscious attempt to realize in any degree the ideals of Robert Owen, or of the Rochdale Pioneers, but has been in the main merely an attempt to procure a safe investment for the surplus capital of the stores, and to satisfy at the same time the craving for a "house of his own" which is so characteristic of the thrifty Lancashire and Yorkshire workman. Similar reasons prompted, in 1884, the formation by co-operators in the South, of the successful society now known as the Co-operative Permanent Building Society, which had in 1906 a share and deposit capital of £333,985, belonging almost entirely to members of industrial co-operative societies, and which, up to the end of 1906, had advanced to its members, on ordinary building society lines, no less than £509,206 upon the security of mortgages on 2134 houses, nearly all of which are occupied by the borrowers, who enjoy the privilege of being the owners of their dwellings.

In 1888, however, a development of the co-operative principle more nearly in accord with the early ideals in regard to housing, was commenced upon the initiative of Mr. Benjamin Jones, then hon. secretary of the southern section of the Co-operative Union. Instead of encouraging workmen to become the owners of their individual dwellings, Mr. Jones urged that the principle of collective ownership would be more advantageous as meeting the special needs of workmen whose work frequently shifted from one district to another; and with the assistance of the late Edward Vansittart Neale and other well-known co-operators, a society was formed in London under the name of The Tenant Co-operators, Ltd. The following summary of the prospectus of this society shows its method of working. Its objects as stated are—

"To acquire or erect substantially built houses, provided with good sanitary and other arrangements for the convenience of tenants. To let the society's houses at fair and usual rents, according to the locality, and variable as the committee deem necessary; to pay a fixed rate of interest on capital (at present 4 per cent.), and to divide all the surplus profits—after providing for expenses, repairs, depreciation, etc.—among the tenants, in proportion to the rents paid by them. Tenants must be holders of at least £1 share in the society, and each tenant's share of profits is credited to him in shares. The advantage to the tenant shareholder is obvious, in that while enjoying security of tenure, the profits are periodically returned to him in proportion to the amount of rent paid by him, and he is not tied to a particular house.

"The advantage to the investing shareholder is that the greater the surplus profits, the greater the security for the regular payment of the fixed interest on capital, while it is to the interest of the tenants who receive the surplus to make these profits as large as possible, *e.g.* by taking care of the property; by helping to find tenants for empty houses; by the punctual payment of rent. Internal repairs are done by, or at the cost of, the tenants, and the share capital of the tenants formed by their accumulated dividends and savings constitute a fund out of which the society, if necessary, can draw to pay for repairs, or to meet any arrears of rent. A tenant remains a tenant only, however large his holding of capital in the society, and if he has to leave the district in which the society's houses are situated, he can sell his shares, or he can continue to hold them and receive the interest upon them."

It is also claimed for this system that, in principle, it solves the question of "unearned increment," for any gain under this head does not go to the shareholders as such, or to the individual tenants in the improving locality, but by swelling the surplus profits it necessarily benefits *all the tenants* of the society, as tenants, in the shape of increased dividends on their rent. The society is managed by an unpaid committee on the lines usually adopted by co-operative stores, including the rule that each shareholder has one vote only, irrespective of the amount of his holding of capital.

The society's houses are situated in five districts, viz. at Penge, Upton Park, Plashet, Camberwell, and Epsom, and includes cottages, double tenement houses, and a block of fourteen flats.

The following table shows the progress of The Tenant Co-operators, Ltd.—

Year	No. of tenements	Total share capital	Loan, stock, and mortgages	Rent received	Net profit including interest on shares and loan stock	Rate of dividend to tenants on rents
		£	£	£	£	s. d.
1889	30	2590	4,818	232	178	1 3 in the £
1891	50	3068	7,633	984	391	2 6 " "
1896	96	4332	14,131	1423	557	2 6 " "
1901	122	4893	20,932	2185	639	1 6 " "
1906	122	5307	20,842	2368	682	0 10½ " "

The example of this pioneer society, for some reason or other, was not followed until about 1901, when, owing to the exertions of Mr. Henry Vivian and others, the Ealing Tenants, Ltd., was formed upon a similar basis, the rules of the Tenant Co-operators being adopted with one or two slight modifications, including a rule that tenants should be required to hold £50 in shares instead of £1. This society has purchased a large estate at Ealing, on which houses of various types have been built, upon the lines now popularized by the Garden City experiment at Letchworth, where also a Garden City Tenants Society has been established. Similar societies have now been formed at Sevenoaks, Leicester, Bournville, and Manchester, mainly owing to the active propaganda carried on by "The Co-partnership Tenants Housing Council," which has now been formed to actively advocate the adoption of this principle of collective ownership in the development of all new estates. None of the societies, with the exception of the pioneer "Tenant Co-operators," have yet paid a dividend to tenants, but there would seem no reason why, when firmly established, they should not all be equally successful. The following table, prepared for the recent Co-operative Congress by Mr. Aneurin Williams, shows the position of each of the Co-partnership Tenants societies formed at the end of 1906.

Name of society	Date formed	No. of members	No. of houses	Share capital	Loan stock	Value of property
				£	£	£
Tenant Co-operators	1888	320	—	5,286	9,030	28,680
Ealing Tenants ...	1901	171	104	8,926	13,935	53,912
Garden City Tenants	1905	111	130	6,126	16,504	37,670
Sevenoaks Tenants	1903	46	47	1,200	3,500	13,500
Leicester Tenants ...	1903	40	—	520	250	—
Bournville Tenants	1906	66	20	2,271	1,165	4,000
Manchester Tenants	1906	30	—	1,060	670	—
	—	784	301	25,389	45,054	137,762

PART III.—THE FARM

CHAPTER XX

THE CLAIMS OF AGRICULTURE

Economic Need of Organization—National Importance of Agriculture—Need of Education

THE application of the principles of association to agriculture is at the present time primarily neither a social nor even an industrial question. It is rather a question of business, the urgency of which may, perhaps, be sufficiently illustrated by the facts that the average price of British wheat from 1894 to 1903 was 27 shillings per quarter as compared with an average of 47 shillings from 1874 to 1883, and that in 1903, out of the astonishing total of 2300 million eggs imported into this country, about four-fifths of a billion came from Russia and Siberia. Not only are changes thus in rapid progress in the conditions of modern trade and commerce that place upon the home markets agricultural commodities drawn from rapidly extending areas, but it is being observed that in a large number of cases these fresh areas derive much of the strength of their economic position from the practice in various forms of those very principles that British agriculturists are only beginning to realize may contain something worthy of their notice.

Thus, the motive that sufficiently explains the degree of attention that co-operative agriculture is obtaining in some quarters is an economic one. It is regarded as a question of business.

Agriculture is, however, a business of an exceptional kind, the successful execution of which fits in perhaps more naturally than does that of any other with a still greater aim than that of business success, namely that of increasing the national welfare. Sound opinion as to the character of every economic act may, it is true, be said to be in the last resource determined by its conformity with this great end, and so interwoven are the economic processes of the nation that it would, perhaps, be difficult to prove that successful development in any one direction was especially advantageous; and it may even be argued in general that not only is agriculture inevitably destined to become an increasingly subordinate interest in this country, but that it matters nothing as to the particular channels by which subsistence and wealth are won.

Even on political grounds, however, such an argument would probably break down, and in its connection with the health of the community, to mention no other point, agriculture easily holds an exceptional position in the industrial hierarchy of the nation.

Moreover, whatever test be applied, agriculture is still seen to be the first industry in the country, and its prosperity has still the greatest indirect influence upon conditions affecting the general welfare. In spite of the immense quantities of foodstuffs imported and destined to be imported into this country, it is still the industry upon which the nation is most dependent. It is the industry that deals with the most valuable, the most permanent, and at the same time the most elastic of our material resources—the fertility of the soil. It is the industry by which, more than by any other, the imagination can be held, and by which affection in exceptional ways can be stimulated. If only the right economic path can be found there are in connection with agriculture many natural instincts to ensure its being trod, over and above the desire for the physical strength that its pursuit tends to give, and the maintenance of physical vigour that it helps to secure. It is, indeed, of those whose occupations are agricultural that it ought, if anywhere, to be reasonable to put the poet's large question—

> "Shall we wake one morn of spring,
> Glad at heart of everything?"

and to hope for a "yes" in answer.

And yet it is this industry, so important, with so many natural advantages, and with so strong a hold upon the traditional imagination of the people of these islands, that is so hard pressed—the comparative economic weakness of which is one of the great causes of the congestion of urban populations and of the declining prosperity of many rural districts.

The weapons by which these ill effects may be lessened or avoided are numerous, pointing, for instance, to the need of fresh applications of scientific teaching, of bolder initiative, sometimes to the need of more capital, more often to its more effective employment, and to other circumstances and qualities—social, moral, economic or administrative—as affecting the individual units of agricultural activity, in this connection pointing, perhaps, above all else to the eternal need of "education," not only of such as will give merely more technical knowledge, but also that kind of mental equipment that enables men to respond with a quick intelligence when new ideas of value are placed before them and to reject those that are merely plausible; that enables them to have greater insight into the economic forces amid which men have to play their parts, and to see something of the bearings of these, working through each individual life, upon the life of the nation.

Experience, however, as we have seen, is also pointing to the pressing necessity of widening out the prevailing conception of what the unit of agricultural activity should often be.

CHAPTER XXI

CO-OPERATIVE AND OTHER AIMS

Slow Progress of Co-operative Experiments — Changes mainly due to Agricultural Community—Individual Initiative most hopeful—Decline in Number of Agricultural Workers

CO-OPERATIVE agriculture has been a perennial subject of discussion from the time of Robert Owen, and from even before the days of the well-known Ralahine experiment in Ireland, so successful and, through the lapse of the man to whom it was due, so sadly ended, something has been always attempted in this direction.

The ends in view have differed, however, ranging from the communistic and comprehensive aims of the early reformers to such a purely business consideration as led to one of the oldest English attempts—the Lincolnshire Farmers' Association, established in 1868 for the simple purpose of purchasing phosphatic manures in large quantities on advantageous terms, and supplying the same to its members at cost price. In a few other cases the primary object has been to unite the interests of labourers and owners, and in these "labour" was sometimes recognized as participant, not only in proceeds, but in management: to some extent the farm, like the workshops of the early Christian Socialists, was to be "self-governing."

Finally, we have somewhat more frequently the alliance of farming, often in the form of market-gardening, with ordinary working-class co-operation. But this side of the movement is not progressive. "We have nothing new to report" is the comment of the Central Board of the Co-operative Union, in its report of 1903.

"About the same number of societies as last year have land which they either own or rent, which is used by them for farming, for market-gardening, and for accommodation for cattle and horses."

In the various sections of the Union the total acreage thus farmed was 7196 by 72 societies, including the two wholesale societies, and £147,511 had been embarked as capital. In something more than half the number of cases, however, losses had been made, amounting in the aggregate to £4225, while aggregate profits amounted only to £3240. Conditions were much the same in 1905, as although by that year the acreage had increased to 8706 and capital to £209,287, the profits had fallen to £4075, while losses were £7889.

In this direction also, therefore, there is nothing either very hopeful or very important to be found, for there is little if any recognition of the alterations that the last twenty-five years have brought about in the economic conditions of agriculture in this country. For this recognition we have to look to the agricultural community itself—that is, to those who, in introducing changes, have been attempting, not to extend the practical sympathy and goodwill of one class to another, and not to democratize an industry, but primarily to adapt the methods of their calling to the changed conditions that have resulted from the widening area of external competition and the rapidly increasing importation of agricultural produce of almost every kind.

Much has been done to secure the necessary adaptations to the altered conditions of the time. Agricultural rents have been reduced;

agricultural machinery and labour-saving appliances are being more widely used; agricultural operations are more intelligently undertaken; the importance of artificial fertilization is being more generally recognized; greater care is being exercised in the selection of seeds and of stock; a new place is being given to what in the past had been regarded as the quite subsidiary portions of farming, while in some parts of the country the "working" farmer has gradually taken the place of those who filled the perhaps more agreeable, but not less important, *rôle* of superintendence; in these and in other ways the process of adaptation has been at work.

It is probably in similar ways, individualistic in their foundations, though often aided, stimulated, and even guided, by various forms of common effort—governmental or voluntary—that the changes will continue to come, and the traditions and habits of agricultural life in England, at any rate, are such that we have probably to look to individual initiative for the chief solution of the very serious difficulties by which British agriculture is still confronted.

Meanwhile the significant increase in the

area of permanent pasture land, the total of nearly 29 million acres in 1905, being over 5 millions more than that of thirty years ago, still tends to continue, albeit slowly, in every part of the kingdom, and is being accompanied by the ominous decline, not simply in the relative but in the absolute number of those engaged in agriculture. In 1871, exclusive of women and girls, the agricultural classes of every description numbered 1,470,442 in England and Wales, and in 1901, 1,094,765. During the same period the males of all ages have been increasing from something over 11 to nearly 15 millions, and it is not surprising, therefore, that by many people the "depopulation of the rural districts" should be regarded as an evil second only in its importance to the correlative evil of "urban congestion."

Thus, the struggle to maintain agriculture in its position of rightful pre-eminence is still in progress, and every practical and hopeful means by which the desired end can be obtained is deserving of special attention and study.

CHAPTER XXII

THE PRINCIPLES OF ASSOCIATION AS APPLIED TO AGRICULTURE

Power of Association—Necessary to Economic Efficiency —Reflex Action upon Character

IF a small English farmer of to-day, who had plodded on in isolated retirement for twenty or thirty years on the lines on which he had been brought up, diligent in his affairs and conservative in his methods, could suddenly realize the agricultural changes that had been in progress during his lifetime, it is not unlikely that he would be above all things impressed by the power of association. If we imagine him to be a man who had wearied of the loneliness and narrowness of his life's enterprise, it might well be that the attractions of associated effort would make him even exaggerate its significance, and find in the social value of the principle something that obscured, not only its own economic usefulness, but also the economic conditions under which alone it could be advantageously introduced. As a rule, this danger is, it is true, very far from being incurred in this country, for it is only after patient propaganda that the rank and file of agriculturists, and then only occasional members of it, have come to see that co-operative enterprise has important applications for them as for others. It is those who are already alive to its value that are sometimes in danger of over-estimating it, and of overlooking the fact that co-operation, if it is to be applied successfully to agriculture, demands just the same personal qualities of assiduity and enterprise for which successful isolated effort also calls.

Far more indeed than in working-class distributive co-operation is this the case. There, the place of the individual in industry is generally almost unconnected with the fact as to whether he is or is not a co-operator : his industrial life is generally something apart. With the agriculturist it is otherwise. Co-operation, for him, is primarily an instrument by which his industrial life may be made more successful, and by which the conditions of that success—greater knowledge, more advantageous buying, more profitable selling, completer mechanical equipment, a sounder financial basis, and enterprise—may be more fully realized. Distributive working-class co-operation is in essence a social principle resting on the adoption of certain industrial practices. Co-operative agriculture, at the present time, is in essence an aid to economic efficiency, accompanied by those moral benefits that always tend to follow on the adoption of associated effort.

Distributive co-operation has to find its chief justification in its effects on life and character; agricultural co-operation, perhaps, in its effects on the stability and expansion of the premier industry of the country. In the former case, business success may be a condition of progress, but it is not the end. In the latter case, the social gain may ultimately be not less than the economic, but it also is not the end. Co-operative agriculture is thus regarded by many as being in essence a mode of business, and it is undeniable that as good business it will have to command the suffrages of practical men and women.

"We do not want in any matter like this," said the late Mr. Hanbury, then President of the Board of Agriculture, at a Congress on Agricultural Co-operation in 1902, "to appeal to sentiment. What is a great deal more useful to appeal to is the principle of pounds, shillings, and pence. That is the principle which ought to guide us whenever we consider how we may advance the interests of agriculture. We should not forget what a very great business agriculture is, and sentiment must be put on one side. We must consider the money results."

And again in an "Appeal to all Patriotic Irishmen," issued three or four years ago on behalf of the Irish Agricultural Organization Society, we read—

"In 630 Co-operative Societies, some 64,000 farmers and other workers of the homestead are carrying on various kinds of production and trade connected with their industry. These associations are, no doubt, business combinations, and give their members the advantages which belong to business organization. But—more important for the nation—they are also centres of business education, where men and women acquire business knowledge in the most effective of all ways, by doing business under competent guidance and advice."

But more than money results and business education have been the inevitable sequel. Sentiment of one kind or another, like cheerfulness in the experience of the pessimistic philosopher, will keep "breaking in."

"The benefits of the movement," writes Sir Horace Plunkett,* "though 'sure' will not come all at once or without great efforts and some sacrifices on the part of those who join it, and are," he continues, "not only of the substantial kind. Every one who has studied the effect of agricultural co-operation all over the world as we have studied it, knows how the mere fact of men meeting together and working together to improve the business of their lives, not only adds to their comfort and prosperity but leads to an all-round social amelioration, and materially adds to the enjoyment of life."

And in the *Report of the Irish Agricultural Organization Society for* 1902,† we find the

* *The Irish Agricultural Organization Society and its Aims.* Leaflet No. 1. of the I.A.O.S., p. 4.

† Page 25.

committee expressing the opinion in the section dealing with Home Industries, that here,

"as in other branches of the Irish Agricultural Organization Society work, the indirect results, the reflex action upon character, are as important as the direct material gain for which the societies are ostensibly organized."

Thus, though the root is found in some form of material advantage, it is necessary to look higher in co-operative agriculture hardly less than in other forms of industrial association for its finer flower and fruit.

CHAPTER XXIII

CO-OPERATIVE AGRICULTURE IN IRELAND

Influence of Systems of Land Tenure—The Transition from Political to Economic Thought and Action—General Figures—Co-operative Dairies—Agricultural Societies—Agricultural Wholesale Society—The Agency—Proposed Federation for the United Kingdom—Co-operative Banks—The Insufficiency of Agriculture—Home Industries—Home Life—The Irish Agricultural Organization Society—Federation—The Maintenance of Local Responsibility—Personal Factors—The Basis of Strength—Ultimate Aims

IRELAND is the pioneer in the United Kingdom in Agricultural Co-operation, and the story of its growth there is beginning to be well known. Following the lead of Denmark in general organization; of Germany and Italy in agricultural banking and credit; but drawing lessons from every likely field, and "at the feet of such men as Vansittart Neale, Thomas Hughes, and George Holyoake," and by constant advice and information from Mr. J. C. Gray of the Co-operative Union, learning also the principles of the co-operative movement in England,—Ireland has been, during the last seventeen years, engaged in a task that seems to bid fair to provide the solvent for some of her greatest ills. Even in the absence of future political changes, the co-operative movement in Ireland has been characterized by economic and moral attributes that under the leadership and inspiration that it had secured, were destined to achieve great things, but at the present time, with a considerable extension of the system of peasant proprietorship taking place, a movement, the key-note of which is organized self-help, should be greatly strengthened.* It should also, at the same time, help

* It does not come within the scope of this volume to endeavour to discuss the question of land tenure, but it may be noted that the greater the prosperity of the individual occupier and the greater his stake in his holding, the greater importance does he attach to security therein. Agricultural co-operation bids fair to increase both this prosperity and this stake, and it will tend, therefore, other things being equal, to flourish most and to take root most easily where the interest of the occupier is greatest—that is, where there is not necessarily ownership, but assured tenure or assured interests in improvements made. These questions are, therefore, often latent in the agricultural co-operative movement. In this connection, the following observation with reference to the recent agricultural revival in Denmark, may be quoted: "The system of land tenure in Denmark, which had done so much to encourage both the creation of agricultural freeholders and the increase of small holdings, had further strengthened the power of the agricultural community to benefit from the opportunities opening out to them." (Pratt: "The Organization of Agriculture" (1904), pp. 29, 30).

to give stability to the industrial basis upon which the financial obligations and thus the political soundness of the great measure of 1903 are dependent.

Ireland is thus a country in which, not only is the need for strengthening the position of agriculture paramount, since over by far the greater part of it, this great industry, subject to much the same stress of competition as in England, is the preponderating—almost the sole—interest, but in which there is also a great suitability for associated effort arising from the prevailing conditions of tenure, and over the greater part, in spite of considerable local variations and of the great problem of congested areas, through the more common size of holding.* Further, there is as compared with Great Britain, a dearth of home markets of great size, and a consequent need that Irish farmers should look beyond the borders of their own land for an outlet for an exceptionally

* It may be noted here that the mere multiplication of small holdings, however brought about and under whatever conditions of occupancy, would often, and especially perhaps in Great Britain, seem to find little economic justification under the competitive conditions of the present time, unless accompanied by the power to adopt the appropriate forms of industrial association.

large proportion of their produce, thus giving alike exceptional scope and creating exceptional need for a machinery of distribution, organized on a large scale.

By far the most important concrete outcome of the movement has been the co-operative dairies, established for the manufacture and sale of butter. Of these no fewer than 269 appear on the register of the Irish Agricultural Organization Society for 1904, and in addition there were 59 "Auxiliaries," that is, branches at which no butter is made, but from which the cream is sent on to central establishments for churning and the later processes of manufacture. The number of auxiliaries thus grouped round a principal dairy varies from one to eight, and although the relation between the two gives rise to a considerable amount of practical difficulty in organization—as regards standard of cream, for instance, and its regular transport—the principles that have led to the formation of well-placed auxiliaries are the sound ones of appropriate localization, division of labour and the economic use of capital, and their numbers, where carting to a central dairy is really impracticable, will probably increase.

Out of the total of 269 dairies mentioned in the report of the Irish Agricultural Organization Society for 1904, a considerable number were not in operation when it was issued, and it is a matter for regret that many others that were at work failed to send in their returns. The following, however, are some of the more important figures that are given: Membership (for 246 societies), 42,432; Paid-up Share Capital (for 217 societies), £112,296; Loan capital, including Bank Overdrafts (for 195 societies), £112,070; Value of land and buildings, after allowing for depreciation (for 198 societies), £212,022; Reserve Fund and Accumulated Profit (for 108 societies), £38,912. The sales of butter, including milk and cream, for 214 societies, amounted to £1,089,620. To 177 societies more than 50½ million gallons of milk were supplied, and by 172 societies something over 20 million pounds of butter were produced. The amount of butter obtained per gallon of milk varied from 6·17 ounces to 8·37 ounces, and the price realized for butter ranged from 9·20*d.* to 12·20*d.* In 116 cases a net profit is shown amounting in the aggregate to £11,156, while in 76 cases the net loss recorded amounted to a total of £5,649.

Net profits, it may be noted, are divided as follows: 5 per cent., as a first charge, goes to members on the amount paid up on their shares; not less than 10 per cent. of what remains to the workers in the dairy in proportion to the wages earned by each, and "the remainder is allotted to the milk-supplying members in proportion to the value of the milk supplied by each." *

It should be observed that figures such as those last quoted, showing financial results, may be misleading, and may be inadequate, as they do not and cannot take account either of the direct and immediate gain to the farmer from the enhanced price at which he is able to sell his produce, estimated to bring him in about £2 more per season on each cow, which is itself equivalent to an advantage of about 25 per cent.; or of the saving of time at the farm owing to the elimination of all the processes of home dairying, save that of milking the cows.

* I.A.O.S. leaflet, No. 3.

In addition to the purely business ends of the dairies which are in themselves comprehensive, and often include those of agricultural supply societies, various other objects are tacked on, not the least valuable being those that are mainly social and educational in character. This side development is indeed characteristic of the whole movement. Reading-rooms and libraries are now a frequent adjunct of the different farmers' societies in Ireland, while "social gatherings of every kind, dances, lectures, concerts, and such-like entertainments, which have the twofold effect of brightening rural life and increasing the attachment of the members to their society are becoming a common feature." *

The table on p. 260 gives further particulars of the growth of these societies.

* Plunkett: "Ireland in the New Century," p. 199.

WORKING OF CO-OPERATIVE DAIRIES IN IRELAND *

Year	No. of societies † furnishing returns	No. of members	Share capital	Loan capital	Paid for milk	Received for sale of butter	Working expenses	Net profit
			£	£	£	£	£	£
1889	1	50	374	—	3,748	4,363	392	223
1894	30	1,641	15,468	6,061	132,555	157,852	15,398	1992
1899	160	22,750	66,346	40,660	484,097	572,963	65,767	3884
1904	246	42,432	112,296	112,070	‡	1,089,620	115,961	5507
1905	248	42,786	114,942	105,737	‡	1,245,486	‡	‡

* Extracted from *Report of the Irish Agricultural Organization Society*, 1905, p. 28.
† The particulars given in the other columns do not cover the whole of these societies, cf. p. 257.
‡ Particulars not published.

Next in importance to the dairies are the agricultural societies, of which there were 151 at the end of December, 1905, formed with the primary object of enabling farmers to save unnecessary intermediate profits in the purchase of seeds, machinery, artificial manures, etc., or, by the strength that association gives, to command the use of more expensive plant, or of more valuable stock for breeding purposes. Many of the societies on the register were not working during 1905, but the returns furnished to the I.A.O.S. from only 88 societies again reflects a blameworthy indifference to the importance of providing the central organization of the movement with complete comparative figures. The total trade of these 88 societies amounted during 1905 to £64,083.

In connection with the agricultural societies, and formed by a combination of about 50 of their number, is the Irish Agricultural Wholesale Society, the transactions of which are now, ten years after its formation, at the rate of about £60,000 a-year.

While the agricultural societies and the Wholesale do something towards supplying the members with what they have to buy, the Irish

Co-operative Agency endeavours to fulfil the not less necessary, but far more difficult, function of helping members to dispose of what they have to sell. At the present time only about 50 of the Dairy Societies are members of the Agency, which was formed in 1893, but, chequered though its career has been and still is, and comparatively small though its operations are, it probably represents a unifying element on the commercial side of the whole movement that in some form or another is not only essential to its growth, but is destined greatly to increase in importance in future years. The special aim of the Agency has been to effect a national combination so that the products of the Irish Creameries may be put, with regularity, with uniformity of standard, and to the best common advantage of the affiliated units, upon the English market. It has been under the auspices of the Agency that the proposal to have a "special national brand or trade mark" has been slowly taking root, by means of which the excellence of the Irish Creamery butter shall secure general recognition among consumers. In 1901 the trade of the Agency amounted to £166,776;

in 1902 to £184,693; in 1903 to £186,701; in 1904 to £169,824; and in 1905 to £166,126.

Up to the present time the Agency has been unable to prevent a considerable amount of "undercutting" of prices among the Irish societies, and the operations of the two Irish Federal Societies have been so far on a small scale. In Great Britain, to which reference will be made in a later chapter, centralized action is in a still earlier stage of development, and so far the advantages possessed by the local societies, either in the way of purchase, production, or sale, have not been so considerable as to attract the larger class of farmer. In order to increase the power of the movement as a whole, in order to prevent internecine competition in the sale of products, and in order to bring the Irish and British movements into line, joint federated action is contemplated.

From these larger projects we may turn again to some of the more modest developments of the movement in Ireland. Not the least useful among these has been the spread of co-operative banking, by means of which credit, hitherto only possible for the small farmer at exorbitant rates of interest, has come within his reach on reasonable terms. The Irish banks have been formed upon the well-known model of the Raiffeisen banks in Germany, with unlimited liability of the members who borrow on the joint responsibility of all, and lend again among themselves at a slightly higher rate of interest.

The usefulness of these banks is receiving exceptional official recognition in Ireland, the Department of Agriculture and the Congested Districts Board having in some cases defrayed the expense of their organization, and in a number of cases advanced a loan for working capital, in non-congested and congested districts respectively, while a very few County Councils have taken the same far-sighted step. The Joint Stock Banks, at first holding aloof, are now coming into line, and the difficulty in securing the small amount of necessary working capital is thus, so far as the credit banks are concerned, slowly disappearing.

As the banks become better known, and as experience demonstrates their solvency, it is hoped that they will be used for the deposit of local savings, and that it will gradually become possible to improve on the general experience of the past, of borrowing at 4 or 5 per cent. and lending at 5 or 6 per cent., and to secure and adopt still easier terms.

The financial basis of unlimited liability makes it a necessary condition that the members, who are generally resident in a single parish, and to whom alone loans are made, should be not only well known to one another, but known also as being honest, thrifty, and of general good repute. Thus—

> "a great educational influence is exercised by the bank in its insistence upon good character, upon proper investments, punctuality of repayment, and by the instruction it gives as to the proper value of money. . . . In these societies more than in any others based on co-operation, is it made evident to all the members—by the fact of their common liability—that in brotherly feeling among them lies their greatest hope of success." *

During 1901–5 the number of these banks increased from 101 to 232, still far fewer than Ireland needs. The Report of the I.A.O.S. for 1906 shows that of the 232 banks in 1905,

* *Co-operative Credit*, I.A.O.S. Leaflet, No. IV., p. 4.

191 only made returns showing an aggregate membership of 13,035, and loan and deposit capital amounting to £38,428. The total amount advanced in loans by the banks was £43,641. The working expenses of the banks amounted to a total of £187 only. The aggregate profit, which is never divided but added to a fund to enable the banks to lend on easier terms, was £531 8*s*. 5*d*.

The dealings of the banks vary greatly both in rapidity and number, but it is a frequent experience for the loans granted in the year to exceed, occasionally more than to double, the total loan capital. The average loans granted also show considerable variations, ranging from the small financing that is reflected in such a tiny average as £1 18*s*. 3*d*. to the more imposing average of £20. For all the banks, however, the average amount of the loans was only £5 17*s*. 1*d*. The amount is small, but "Labourers' Banks" may in the future tend to make it a still more modest one. One of these banks only was registered in 1902, to enable this class—which will need fuller recognition than it has yet been able to receive, either in Ireland or in Great Britain, especially in the districts of larger farms—to take a greater part in some of the auxiliary occupations of agriculture, such as the keeping of pigs, poultry, and bees.

The figures are still small, but their social and industrial significance is very far from being expressed in terms of currency.

Supremely important though agriculture is in Ireland, and numerous though the projected developments of co-operative enterprise are—winter dairying, flax growing, poultry fattening, fruit and vegetable and flower growing, jam making, cider making, to name some only among the number—it is still felt that however intensive, and varied, and even successful agriculture may be, it can never support a growing population, or alone check emigration. Other industries must also be supported and followed.

In consequence of the urgency of this need, the economic basis of the whole agricultural movement has indeed been adversely criticized and its inadequacy proclaimed. The facts of the situation are, however, appreciated by the leaders of the agricultural movement itself, and with apparent inappropriateness we find among its offshoots, in addition to urban co-operative societies—for the production of boots as at Ballina, and of shirts as at Sligo—various rural societies for the promotion of home industries. For these which "engage in such industries as embroidery, sprigging, weaving, knitting, and shirt-making," and produce a variety of articles including among others various kinds of lace, drawn-thread work, vestments, underclothing, hats, rugs, and dolls, it is still the day of very small things. Available returns are in this group even unusually incomplete, and in 1905, out of 50 societies, only 28 furnished particulars, showing a trade for the year of £13,018. In 1901 the total trade recorded for about 30 societies amounted to £8950. In 1902 the corresponding total for 32 societies was £11,998.

In connection with this work it is observed that the difficulty in organizing it is, not to obtain workers, "for the desire for industrial development has grown rapidly over the country," but workers possessing the requisite technical skill or manual training, or, for several branches of the work, those who have been sufficiently trained in drawing and design. It has also been rather exceptionally difficult to obtain satisfactory and expanding markets for the articles that are made, and in too many instances operations have resulted in financial loss.

In spite of the comparative commercial insignificance of these attempts, it has been admitted by the Committee of the Irish Agricultural Organization Society that no branch of its work is more important.* It is hoped that through these industries much may be done to provide alternative employment to those who under the old system were occupied at home in dairying processes that are now undertaken by the factory; that at the same time something may be effected to prepare the way, not perhaps for a great industrial revival in Ireland, but for some rehabilitation of its lost industrial life and habits; and, further, that their pursuit may react beneficially upon Irish home life.

If the lurking danger, lest the supplementary home worker fall too low in the industrial market, be avoided, there is everything to be said for the reasonableness and appropriateness of this home industries movement, calculated

* See *Report*, 1902, p. 23.

as it is to give employment to the leisured or partly leisured members of the farming families, for whom often no other outlet than emigration offered, and to add interest and variety to village life. In addition to the improvement in the home that may result from increased incomings, an attempt is also being made in Ireland to associate the renewal of industrial life with the amelioration and beautifying of home conditions themselves. To some extent this gain comes as a matter of course, since all work that has in itself some æsthetic quality tends in the absence of gross economic pressure to react upon the conditions of life under which that work can best be conducted, and thus—

"the disorder, slovenliness, and squalor, too often met with in the Irish homestead, and hitherto accepted as a matter of course, become recognized as defects capable of remedy, and the desire for neatness, order, and cleanliness insisted upon in the workroom extends to the home." *

Thus the growth of the Irish agricultural co-operative movement has not been sporadic,

* See *Report*, 1902, p. 25.

but has been accompanied by a degree of co-ordination and a width of view that ought to afford a safeguard for continued expansion. To a great extent not only the growth itself, but the fruitful way in which it has taken place, has been due to the magnificent work achieved by the Irish Agricultural Organization Society; but although so great a debt is owed by Irish agriculturists to this society, it has consistently avoided the assumption of a burden that the farmers themselves should bear: propaganda and advice have been kept rigidly distinct from the actual promotion of business undertakings, and the democratic, self-reliant basis of every affiliated unit of co-operative enterprise has been preserved, as furnishing the only sure foundation on which the movement can stand. This cardinal feature has been sometimes overlooked both in Ireland and in Great Britain; but the position that experience now shows to be the safest is put with great clearness and directness in the *Report* of the Society for 1902:—

"The committee desire to give expression here to their unqualified disapproval of any scheme, however well-intentioned, which is designed to transfer from the Irish farmer to an outside body any form of business which the farmer is capable of doing himself by association with his fellows. They hold that such schemes can only do harm by weakening the spirit of self-reliance which it has been the great object of the Irish Agricultural Organization Society to engender." *

This policy brings with it some characteristic dangers, such as absence of uniformity in quality of product, in standard of financial control, and in technical management; and in the avoidance of such dangers much remains to be accomplished. Many of the necessary steps are, however, in train, partly through the influence of the Irish Agricultural Organization Society—the government of which has been placed in 1904 upon a more representative, and, to that extent, upon a more responsible basis—partly through the gradual adoption of the federal principle, and partly through the inherent educational influences of the movement itself.

The same principle of local responsibility

* Page 32.

has so far been observed in the steps taken towards federation—steps that are themselves the inevitable sequel to the multiplication of the constituent units of co-operative enterprise.

"Every year," says the *Report* for 1902, "emphasizes more and more forcibly the fact that in this age isolated action by societies is almost as impotent as isolated action by an individual." *

But the foundations of the movement towards greater centralization must, it is seen, be well and truly laid: federation "must be built up from the bottom." It

"must be a gradual and, possibly, a slow development; its necessity must be fully realized by the local bodies, which must be educated to regard it as a means whereby their work may be rendered more perfect and efficient, rather than as a refuge from responsibilities and work which as co-operators they must discharge. In short, federation must only supplement—not supersede—local efforts, and in the constitution of the scheme the obligations of each and every society must be clearly laid down and accepted." †

* Page 29. † Page 30.

The hope may be expressed that these opinions will be kept carefully in view, as the new federal scheme mentioned on p. 263 is planned, and the policy of the past be adhered to at all costs, even at that of a slower growth, to avoid the creation of a co-operative bureaucracy. For even in such an industrial and democratic movement as that of co-operation, no matter what constitutional safeguards may be devised, bureaucracies are ever tempted to assume power and responsibility that should be shared by others, and that, if not thus shared, tend in the long run to hinder the growth of that very movement which the bureaucracy itself may quite honestly desire to help and to foster.

In this Irish co-operative movement, however, it happens that propaganda and personal effort, although not official control of any description, have played so large a part that the movement itself is very closely associated in the public mind with the name of one man, to whom it is quite true, far more than to any other individual, its success is due. In the English movement, save for a few historic names, it is otherwise, and its organizations, large and small, owe their vitality, by general consent, to the efforts of very large numbers of the people: they flourish, for the most part, under a democratic anonymity.

In Ireland, however, as in England, there has been, and is, although not to the same extent, the root from which success has sprung, made up of the fibres of many lives, and Sir Horace Plunkett himself is probably more conscious of this than any other person, knowing better than any other on how insecure a basis the movement would otherwise be resting, and that inevitable failure, however great the temporary success may be, would await it. Thus we find him drawing attention to this fact in his volume on "Ireland in the New Century," and while he mentions a few of those whose services have been, and are, pre-eminent in the work — among the number Lord Monteagle, "a practical philanthropist if ever there was one"; Father Finlay, Mr. R. A. Anderson, the present Secretary of the Society, and Sir Horace Plunkett's "chief fellow-worker in the early struggle," and the two Assistant Secretaries, Mr. George W. Russell and Mr. P. J. Hannon

—he writes also of the "unknown leaders," paying his tribute

"of respect and gratitude to those true patriots who have borne the daily burden of the work. . . . By these men and women," he writes, "applause or even recognition was not expected or desired: they knew it was to those who had the advantage of leisure, and what the world calls position, that the credit for their work would be given. . . . As Irish leaders, they have preferred to remain obscure, conscious that the most damaging criticism which could be applied to their work would be that it depended on their own personal qualities or acts for its permanent utility."

They have built their hopes of human progress upon human character, and, as such, rank among "the real conquerors of the world."* Such, indeed, is the condition of every stable movement — religious, political, social, or industrial—that it should rest on this great foundation and be thus permeated by the influence of "unknown leaders," but the truth is sometimes overlooked by those

* "Ireland in the New Century," p. 209.

whom the world knows, and it rarely finds such simple and whole-hearted expression as it has done in Sir Horace Plunkett's book.

Thus the Irish movement definitely places before itself a far more ultimate goal than that of insuring commercial success, and adherents and onlookers alike are warned against seeing in it nothing but the material ends exemplified by "creameries and the cheap purchase of manures."

Financial success is to be the framework only upon which the structure of greatest value is to be reared of a happier, a more self-reliant, and at the same time a more social life.

On the other hand, the danger is avoided of belittling, in any way, the far-reaching significance of industry itself, and the due fulfilment of its tasks is recognized, not only as adding to the happiness and dignity of the individual life, but as forming the foundation of the prosperity and well-being of the whole community.

Greater material comfort, intellectual and moral advance, and the progress of a nation are thus the three great ends of this admirable new Irish movement.

CHAPTER XXIV

CO-OPERATIVE AGRICULTURE IN GREAT BRITAIN

Early Stages—The Agricultural Organization Society—Agricultural Supply and other Societies—The National Poultry Organization Society

WHEN we turn to Great Britain we find that agricultural co-operation is here still almost in its infancy. As already mentioned, English co-operative experiments in this field have been made from time to time for many decades, but what may be called the new movement, based upon the principle of the voluntary association of local producers for specific purposes, be it for buying and selling (as in agricultural societies), for making (as in dairy work), for financing (as by agricultural banks), for collective insurance, or for securing the various advantages that federation and the possession of a central propagandist and advisory organization can secure, Great Britain is but entering upon the road along which Ireland has already travelled so far.

The focussing point here is now the Agricultural Organization Society,* founded in 1901 and based avowedly upon the model of the corresponding Irish body. The object of the Society is comprehensively stated in the following terms:—

"To secure the co-operation of all connected with land, whether as owners, occupiers or labourers, and to promote the formation of agricultural co-operative societies for the purchase of requisites, for the sale of produce, for agricultural credit, banking and insurance, and for all other forms of co-operation for the benefit of agriculture."

As in Ireland, religious and political questions, and the advancement of any political party or religious body are eschewed; and also, as in Ireland, the society is non-trading and solely propagandist and advisory. It sends out organizers to speak or give advice; it provides model rules; it arranges for lectures;

* Until 1905 known as the Agricultural Organization Society of Great Britain. In that year a separate society for Scotland was formed.

it acts as an information bureau to affiliated societies; it furthers combined action between the societies in every way possible for the advancement of the common trading interests; and it publishes leaflets and pamphlets,* and issues numerous circulars dealing with the various forms of agricultural co-operation. Its propaganda has already attracted considerable attention, and in this fact an indication may be found that its objects are well chosen, and in accordance with changes towards which the country is so far rather blindly feeling its way, and concerning which it needs guidance. Up to the present time the formally constituted Societies, although increasing in number rather rapidly, are not very numerous, and with few exceptions derive their importance rather from their significance as types of what may be attempted, than from the actual extent of their transactions.

* Among these are the following: *Hints on the Formation of Co-operative Agricultural Societies; Hints to the Secretary of an Affiliated Society; Model Rules; Agricultural Banks; An English Co-operative Dairy at Work; Feeding Cake Trade; County Councils and Agricultural Co-operation; Extracts from the Fertilizers and Feeding Stuffs Acts;* and *Hints on the Keeping of Books used by Co-operative Agricultural Societies.*

Agricultural supply societies make up a great majority of the affiliated bodies, accounting for 96 out of a total of 134 that were on the register at the end of June, 1906. In addition there were at that time 12 dairy societies, 4 allotment societies, 11 credit societies, 3 motor service societies, and 8 miscellaneous societies.*

The following figures, extracted from the last report, will give future information as to recent progress:

	1901	1902	1903	1904	1905
Total number of societies	25	41	72	98	123
Total membership	517	1094	3245	4926	7439
	£	£	£	£	£
Total turnover ...	9467	16,274	38,909	136,677	221,524

The activity of the movement is thus apt to be rather dependent upon the presence in a district of a few enthusiastic and effective supporters or propagandists ; as have, for instance, been forthcoming in South Wales,

* Including "The Agricultural Co-operative Federation, Ltd.," formed in 1905 to supply the requirements of the affiliated agricultural societies.

in Gloucestershire, in Sussex, or in Yorkshire, but it is worthy of note that the Society is now, in some measure, represented in no fewer than 41 counties, as compared with only eight in December, 1901.

Working in a much narrower field than the Agricultural Organization Society is the "National Poultry Organization Society," a body founded about two years earlier, in 1899, with the primary objects "of encouraging and developing the production of the best qualities of poultry and eggs in the United Kingdom," and of securing, by locally combined efforts, better facilities in agricultural districts for marketing these products. The salient features of the scheme of operations are to form "branches" whenever the requisite local co-operation can be obtained, and, in suitable districts, "collecting depôts."

The depôts are formed for the collection of both eggs and poultry, but, so far, in the great majority of cases they handle the former product only. The eggs are collected, tested, graded, and those that are "strictly fresh," having been branded with the trade mark of the society, are packed and forwarded as rapidly as possible to traders whose names are on the register kept by the central organization.

The various branches (8) and depots (19), "poultry societies," etc., that had been affiliated up to June, 1904, numbered 38. Particulars of membership or shareholders (not all of whom are "producers"), are given in 28 cases, and the numbers range from 8 to 350. The average is about 60. The price realized for eggs per dozen ranges from 1*s.* to 1*s.* 3*d.*, and in only a single case is 1*s.* 1½*d.* exceeded—the uniformity of average illustrating in an interesting way the fact that when once the machinery is provided for keeping in touch with considerable centres of consumption, the opportunity for great local variations such as exist still in many unorganized parts of the country tend to disappear.

The uniformity of price also demonstrates the effectiveness of the help given locally from head-quarters, the aim being so to guide distribution that overlapping and competition in the same markets by the local depôts may be avoided. For this purpose a manager, in touch with traders all over the country, operates from London, and, within the limits imposed by distance and cost of carriage, is thus able to direct supplies as conditions of demand and price determine.

It is estimated that the total value of eggs and poultry consumed in Great Britain in 1903 was more than £17,000,000, and as the opinion of the society is probably well-founded that, although nearly half that is consumed was imported from foreign countries, "all the foreign supplies could have been produced in the United Kingdom without the displacement of other agricultural products," it follows that as the society makes its influence felt, one very important effect will be, not only to increase present values, but to add to the national produce, since the vast and increasing demand of the country may be relied upon to make itself effectively felt, as the possibilities of poultry-rearing and egg-production are improved and appreciated.

For the educational and perhaps for the commercial activity of the Society, there remains a wide field, but the very exceptional and contracted circumstances under which it seems possible for poultry farming to be successful as a separate industry, make it doubtful whether on more strictly co-operative lines than the society has yet attempted, much could be accomplished. "Poultry farming" is, indeed, a misleading term, tending, as it does, to raise hopes that are very unlikely to be realized, and to conceal the fact that to be successful, poultry-rearing and egg-production have in the vast majority of cases to be regarded as subsidiary occupations.

CHAPTER XXV

OFFICIAL PROPAGANDA

Attitude of the Board of Agriculture — Extracts from Pamphlet on " Farmers' Co-operative Societies "—The *Boerenbond* in Belgium

IN Ireland the close and sympathetic connection existing between the new Department of Agriculture and Technical Instruction—that admirable example of constructive statesmanship—and the co-operative movement is well known, and in Great Britain on the part of the Board of Agriculture there has been, especially during and since the presidency of the late Mr. Hanbury, a corresponding cordiality. It would be disastrous if any form of state assistance to such a movement as that of co-operation were to depart from the principle laid down by Sir Horace Plunkett, that such aid "must be preceded, or at least accompanied by, the voluntary association of those to whom the aid is to be given ;" * but the markedly sympathetic attitude of a department of state should in itself impress

* *Report of the I.A.O.S.*, 1902, p. 2 .

upon the farmers of Great Britain the fact that the principles of co-operative agriculture rest upon no insecure foundation. The following extracts are taken, not, as might have been suspected, from a leaflet issued by the Agricultural Organization Society, or by any kindred propagandist body, but from an eight-page pamphlet on *Farmers' Co-operative Societies*, issued, in December 1903, gratis by the Board of Agriculture.

"Next to the [co-operative credit] banks" the most common, and perhaps the most effective form of combination amongst farmers is to be found in the joint purchase societies, or agricultural trading associations. Their usual function is to purchase *wholesale*, manures, feeding stuffs, seeds, implements, and other articles used on the farm. By purchasing in large quantities direct from the manufacturer, these societies are able to obtain supplies for their members at wholesale prices.

. . . . .

"Some of the agricultural trading societies are also able to assist in the improvement of the live stock kept by small farmers by purchasing or hiring first-class bulls, boars, and stallions.

. . . . .

"Co-operation in production has been applied with greatest success to the dairy industry. The remarkable development of the butter trade of Denmark is attributed largely to the establishment of co-operative dairies and creameries, which have enabled the farmers of that country to supply the British market with immense quantities of butter of uniform quality. Uniformity in flavour, in appearance, and in consistency, is the characteristic most required in butter intended for general consumption in the great towns of this and other countries; and it is obvious that this is more likely to be secured by manufacturing the article in dairies which can manipulate the milk supplied by a large number of farmers, than if each of these farmers himself makes butter from the milk produced on his own farm.

. . . . .

"They [the co-operative poultry societies] purchase eggs as well as poultry from their members *by weight*, and the introduction of this practice is said to have had the effect of making poultry-keepers more interested than before in maintaining a good breed of fowls.

"In every case the price for the eggs sold through the societies has been above that obtained before they were started.

. . . . .

"Many of the complaints made by farmers of excessive and preferential railway charges arise from the fact that the consignments concerned are not sufficient in bulk to enable the companies to handle them with profit at the lower charges at which they convey larger consignments. In such cases the remedy would frequently be found in the formation of a co-operative distributing agency, which would undertake the collecting and packing of small consignments to make up trucks-loads for dispatch at regular intervals."

The various examples of association given in the pamphlet are held to be "sufficient to afford some idea of the directions in which farmers, and particularly occupiers of small holdings, can effectively combine to their mutual advantage." In conclusion, it is pointed out that the efforts of propaganda, such as those of the Agricultural Organization Society—

"can now be seconded by County Councils in rural districts where co-operation is likely to be useful, the Board of Education being prepared to sanction * the teaching of the principles and practice of agricultural co-operation in the case of all County Councils which

* This sanction has been given in many counties.

may make application to them in terms of Section 8 of the Technical Instruction Act, 1889, provided the Board are satisfied that such a form of instruction is required by the circumstances of the district."

The extracts reflect the tenor of a pamphlet which is significant of the official point of view, and its publication is a further proof that the principle of industrial association is one that the farmers of this country, especially the smaller ones, can no longer afford to ignore.*

The exact steps taken to give it effect will necessarily vary with local conditions and opportunities, according, for instance, to the nature of the soil and crops, to the basis of tenure and size of holding, to the existing facilities for transport and sale, and for the present until the movement "catches on," also

* A similar pamphlet was issued by the Board in June, 1904, explaining the advantages and working of "Co-operative Egg and Poultry Societies;" drawing attention to the specially suitable field here for associated enterprise because of the perishable nature of the products with which these societies have to deal, and of the facilities for rapid marketing which are therefore necessary, and which the societies can provide; giving examples of success both in England and Ireland; and referring readers, for the supply of model rules, etc., to the National Poultry Organization Society.

with the scattered energy that personal enthusiasm and personal knowledge will bring. At least a beginning has been made, and even with us, as the advantages of one application after another of the new system become known, the spark may at any time become a flame.*

* It is as improbable as it would be undesirable that the movement in this country should be identified with any party, or with any sectional propaganda, be it, for instance, to strengthen a Church or to weaken Socialism, as has happened, for instance, to a great extent in Belgium and Italy. But the following graphic illustration of rapid development in the former country may well find some healthy counterpart with us :

"A Flemish farmer at Goor went one day to the curé of the parish, M. l'Abbé Mellaerts, and spoke to him about the poor quality of his wheat crop. The curé had studied botany and kindred subjects at his seminary ; he had especially followed up the subject of chemical manures, and he had made experiments on his own account in the garden of his house. So he asked the farmer, 'If I tell you of a remedy, will you use it?' 'If it is not too dear,' was the reply. When the farmer called again the abbé gave him a sack containing 25 kilogrammes of chemical manure. The farmer was reluctant to take it. He had no confidence in such manure as that because it did not smell strong enough. But he was induced to try it as an experiment, and he used it to grow some potatoes, with such excellent results that he went to the curé for more. Then several of his neighbours wanted supplies as well. Meanwhile the curé had been reading of what the peasants along the Rhine had done in the way of forming combinations for the joint purchase of agricultural necessaries, and he called a conference of members of his flock to consider the adoption of a like scheme for Goor. His parishioners had no great faith in the proposal, but seven of them put their names down as members of a 'Peasants' Guild'—just to please him. They soon found, however, that they could get their supplies cheaper and of a better quality through the Guild than they could individually, and thereupon more members joined. Within a year the Guild consisted of 100 farmers. Considerations of health then compelled M. Mellaerts to remove to Louvain, where he became an active writer on agricultural questions, and an especially earnest advocate of agricultural combination. A conference of agriculturists at Louvain, organized by M. Mellaerts and others, followed in July, 1890, when it was decided that there ought to be in every commune in the province an agricultural association similar to the one at Goor, and that when formed, all of them should be connected with one central body. By the following year there were 89 local associations of different kinds ready for incorporation into an organization to which the name of 'Boerenbond' was given. By 1893 the number of affiliated associations in the federation was 130. In 1897 the total increased to 380, and in 1900 to 450, representing upward of 26,000 members, and covering the provinces of Antwerp, Brabant, and Limbourg. The federation publishes a monthly agricultural review, holds innumerable conferences and periodical meetings, conducts experimental fields, has a central office from which a vast amount of gratuitous practical advice is given, exercises a useful influence in regard to legislation affecting agriculture, and carries on so big a business in grouping the orders of the local associations that it has organized a separate section for each commodity, set up a mill of its own for the preparation of feeding-stuffs, and established a wholesale warehouse of substantial proportions in the city of Antwerp—all this being done in little more than a dozen years. To the original founder of this great federation is further due the introduction and popularization in Belgium of Raiffeisen agricultural credit banks, of which there are close on 200, with about 10,000 members, in direct connection with the Boerenbond alone."—(Pratt's "Organization of Agriculture," pp. 96–99).

Still more striking, perhaps, is the fact that in Holland co-operative dairies increased from 19 in 1900 to 539 in 1902, with 40,000 farmers then estimated to be sending in their milk and producing butter to the extent of nearly 15,000 tons in the year. (See Pratt, *op. cit.* pp. 136–7.)

CHAPTER XXVI

A CO-OPERATIVE OPPORTUNITY

Economic Reasons for Co-operative Dairy Farming—The Milk Supply—A Question of Hygiene

AS in the case of so many other countries, it is probable that dairy-farming will also in England gain most from the application of co-operative principles, because it is here that not only is there considerable scope for improvements in the purely farming processes involved, and for economies in transport and in marketing, but it is also here that modern invention gives greatest scope for the transition from the home to the factory system of production—from the "small" to the "large"—a transition in which is, perhaps, seen the most distinctively economic feature of the new movement.

In a country like Great Britain, however, with its large centres of population, it is not improbable that in many grazing districts it is in the preparation and sale of milk rather than in that of butter or cheese, that the co-operative principle will be found of exceptional use.

Apart from new laid eggs and fresh vegetables, and these only of certain kinds, milk is almost the only agricultural product of which local producers can maintain a monopoly, and even as regards milk, the "locality," in the case of great centres of consumption, covers a radius of many scores of miles. But although milk may be, and often is, brought long distances by rail, it cannot well be imported, and the fresh milk that England consumes England will have to supply. No article of food is more important, none perhaps is so liable to adulteration and impurity, none ought to be more generally consumed, and none in the interests of the health of the community ought to be more free from suspicion. It is in the supply of this commodity, therefore, that an exceptional opportunity seems to exist at the present time for co-operative enterprise. The most authoritative opinion available of bacteriologists and doctors might with advantage be obtained by its leaders as to what form of sterilizing or pasteurizing is really hygienic, or, as is probably necessary, a special investigation and report might be instigated, and an attempt made not only to spread the knowledge among the public as to the importance, especially for children, of securing a supply of milk, the purity of which is guaranteed, but to justify the belief that such a supply is assured when it comes from co-operative sources.

The reasonableness in this attempt is of course dependent not only upon the fact that the larger the enterprise the easier it is for management and inspection, to safeguard every step in production and sale, but also upon the conditions of a satisfactory supply, associated effort making it possible to secure the equipment of the somewhat costly machinery and plant for "separating" (which is primarily a process of cleaning) and purifying the milk that the isolated producer, save in exceptional cases, cannot hope to provide.

Special attention has been drawn to the need in the interests of the health of the community of a more abundant and purer milk supply, and to the opportunities that this need seems to afford for co-operation. The economic importance and possible future of this minor field of the great industry of agriculture is not so inconsiderable as it may at first sight appear,* but in many other directions to which

* In the valuable Report on the Production and Consumption of Meat and Milk in the United Kingdom, promoted by a committee of the Royal Statistical Society (see *Journal*, Sept., 1904), the average yearly consumption of milk per head for the whole country and for all classes is put at 15 gallons, or a little less than a quarter of a pint per person per day. Labourers (mainly agricultural) are credited with 5 gallons per head, and artisans, mechanics, and labourers (mainly urban), with 12 gallons.

In this Report the total production of the United Kingdom consumed as milk is put at 620,000,000 gallons. If we take the unweighted average of lower middle, middle, and upper class consumption, which, as given in the Report, works out at just under 32 gallons per head per annum (or about half a pint per day) as what *ought* to be consumed, we have a shortage in consumption (and in supply) amounting to about 700,000,000 gallons.

The valuable Board of Trade returns (Cd. 2337—1904) give particulars of family expenditure for 1944 urban workmen's families, and in these, the family representing an average of 5·6 persons, the average consumption of fresh milk is only 9·91 pints per week, equivalent to little more than a quarter of a pint per person per day. The analysis of the returns by districts, assuming that the size of the family is uniform, shows variations from nearly two-fifths of a pint per person per day in Scotland—still, relatively, the land of porridge-eaters—to about one-sixth of a pint in London. Even such small quantities are probably very considerably above the general average consumption in working-class families throughout the country, for, as is pointed out in the Board of Trade returns, "the higher range of family incomes is unduly represented."

It may be noted that in 1903 we imported considerably less than three pounds per head of "condensed," that unsatisfactory reference has been made the principle of association is probably destined, even in this country, to find far larger, if not more useful scope.

substitute for fresh, milk, together with a negligible quantity of fresh or preserved milk and cream.

CHAPTER XXVII

RAILWAY RATES—THE PROBLEM OF DISPLACEMENT

Railway Rates—The Problem of Displacement—Conclusion

TO the vexed subsidiary question of railway rates, which is felt by many to underlie all projects for the revival of agriculture in this country, only incidental reference has so far been made, but because of its almost universal importance an additional word must be written.

Haulage, it may be noted, is probably destined, perhaps in the near future, to be one of the ways in which co-operative enterprise will manifest itself, and co-operative lessons will doubtless be learnt, for instance, from the experimental service of motor-lorries that the North Eastern Railway Company, in close association with one of the most flourishing centres of agricultural co-operation, has already started in Yorkshire for the collection and distribution of heavy traffic in remote agricultural districts. For the present, however, if not within carting distance of his market, practically every farmer in the country is dependent more or less upon existing railway transport, alike for what he buys and what he sells.

As to the extent, however, to which he is helped or hindered by the railway companies, the most conflicting opinions are held, and while some consider that the companies are a standing obstacle to every useful project, others maintain that their policy is one of great consideration and enlightenment, and that the fault, when fault there is, lies rather with the agricultural classes themselves.

The fact appears to be that responsibility in this matter, as in so many others, has to be divided, and that while excessive and restrictive rates are not infrequently charged and vexatious regulations often made, it is also the case that, often through lack of association, the facilities offered by the companies for cheap or convenient transport are not used.*

Thus, on the side of the farmers it has,

* Cf. Pratt: "The Organization of Agriculture" (1904), ch. xxv.

perhaps, above all else, to be recognized more generally that it is unreasonable to expect to have small lots carried at the same rates as large ones, and that by combination it would be often possible to save considerable sums by arranging for the despatch and receipt of larger and more regular consignments.

On the side of the railways the economic truth that prosperity breeds prosperity has to be still further realized and acted upon, since facilities for transport that help on the recovery or development of an agricultural district react in diverse ways, in some of which the companies would themselves inevitably share. Home producers ought at least to have good grounds for expecting to be safeguarded against anything of the nature of preferential rates in favour of importers by this consideration, for in the one case the rate earned is the beginning and end of the railway transaction; in the other it is only the beginning. In the one case the money paid for the goods consigned is spent, it may be, the other side of Europe, and in the other in all probability in the very villages and towns served by the railway company itself. In articles requisitioned for the farm and the home, in holiday junketings, and in many of the various ways in which the earnings of prosperity percolate through a district, the railway company would be sure to secure its toll.

The same argument, based on the transmission and difficulties of prosperity, may be, to some extent, hopefully applied in a quite different field, namely in the case of intermediaries such as agents, dealers, and private traders, who find themselves displaced by co-operative organization, and their services no longer required. In some cases, undoubtedly, such displacement would lead to hardship, and this result, in the postponement or avoidance of which some compensation may often be found when progress seems slow, also furnishes the inevitable subject for regret in almost every economic re-adjustment. Even if those displaced deserve as individuals scant pity, such as the notorious "gombeen" men of Ireland, it does no harm to remember that such men in all probability have wives and children.

Usurious dealing is, however, a transaction of an exceptional kind. In general, business is reputable. Fairness is more prevalent than unfairness, and for men who, having pursued perfectly honourable business careers, only to find their past functions useless, the new system would often be able to provide new duties for the fulfilment of which their technical and trade knowledge would admirably qualify them.

Even if those thus displaced did not or could not join the ranks of the new co-operative executive that would be required—as managers, organizers, instructors, secretaries, or representatives at market centres—the success of the very movement that had displaced them might still provide, directly or indirectly, the fresh fields that they required.

It is in any case inexpedient, and perhaps also somewhat pharisaic, in spite of all that may be true that is said of "rings" and attempts to boycott, of sharp practices and shady dealing, to consider that co-operative agriculture has it in any sense as its primary or even as a very important part of its task "to work against fraud and injustice" (to quote the words of a propagandist leaflet), or to enable farmers to combat traders who are "constantly imposing" upon them.

Industrial association is, indeed, in the first place not a struggle against unfairness of any kind, but rather, when it is the appropriate instrument, the adoption of a new and potentially a better system than the old one, and one that should justify itself moreover not only by its economic strength, but also by the more social and more assured basis upon which it would help to place the everyday industrial life of the nation.

The characteristic effect of prosperity in the organized life of modern society, in spite of the numerous ganglia of over-congestion that it displays, is diffusion, and industrial co-operation is in essence not an attack or even a stricture upon a class, but a stage in industrial evolution, in the advantages of every healthy manifestation of which all classes, and all neighbourhoods—agricultural labourers, for instance, no less than the farmers and landowners, the towns not less than the country districts—should ultimately in some ways share.

INDEX